Weinert   Plastic Optical Fibers

# Plastic Optical Fibers

Principles
Components
Installation

By Andreas Weinert

*Mit freundlichen Gruß*

*Sep. 2005*

Publicis MCD Verlag

Die Deutsche Bibliothek – CIP-Einheitsaufnahme

**Weinert, Andreas:**
Plastic optical fibers : principles, components, installation / by
Andreas Weinert. [Ed.: Siemens-Aktiengesellschaft, Berlin und
Munich]. – Erlangen ; München : Publicis-MCD-Verl., 1999
    Dt. Ausg. u. d. T.: Weinert, Andreas: Kunststofflichtwellenleiter
    ISBN 3-89578-135-5

ISBN 3-89578-135-5

Editor: Siemens Aktiengesellschaft, Berlin and Munich
Publisher: Publicis MCD Verlag, Erlangen and Munich
© 1999 by Publicis MCD Werbeagentur GmbH, Munich

Printed in Germany

# Foreword

No matter what form of information is involved – data, voice or images – the quantities of data to be transmitted are increasing all the time in every sector. Fiber optic technology, which is already widely used in the telecommunications sector today, is now also emerging as a cost-efficient and secure transmission medium over short distances within or between buildings.

Plastic optical fiber is a very new transmission medium and is being used today not only for data transmission, but also in numerous other applications, such as lighting and sensor technology. In recent years the overall system for data transmission with plastic optical fibers has been developed to such an extent that the connection of a plastic optical fiber to a connector is simpler than, for example, that of a shielded copper cable. For this reason, there has been a tremendous rise in the installed number of transmission routes using plastic optical fibers. Especially in the field of industrial automation sector, both inside machines and externally, plastic optical fibers are being put to greater use.

This book deals exclusively with the use of plastic optical fibers for data transmission. Apart from the physical principles and the manufacturing process of the fibers, the individual components of transmission route (cables, transmitter and receiver components) as well as the necessary connector technology are also covered. In addition, important installation instructions are provided, as well as a preview of possible future developments. Major national and international regulations are specified and sometimes explained at the appropriate points.

This book is intended for engineers, technicians and developers as well as users involved in the installation of plastic optical fibers. As a result of its detailed description of the principles, the book is also suitable for students and staff of technical colleges and universities. As plastic optical fibers can be processed without costly tools, simple experiments to aid the understanding of plastic optical fiber technology can be implemented at no great expense.

This book lays no claim to being an exhaustive scientific publication, but will give the reader a clear and comprehensive insight into fiber optic technology in general and plastic optical fibers in particular.

Neustadt b. Coburg, November 1999　　　　　　　　Andreas Weinert

# Contents

# 1 Physical fundamentals

## 1.1 Principle of optical signal transmission

The principle of optical signal transmission is shown in Fig. 1.1. A transmitter component converts the electrical signal into an optical signal and launches it into the optical waveguide (optical fiber). The optical signal then travels along the optical fiber until it is converted back into an electrical signal by a receiver.

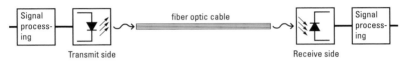

**Fig. 1.1** Schematic principle of optical signal transmission

Where necessary, the term "optical fiber" is abbreviated in this book to OF, "plastic optical fiber" to POF and optical waveguide to "OWG".

We will first consider the individual physical variables and effects that are necessary for an understanding of optical signal transmission.

## 1.2 Light propagation in optical fibers

### 1.2.1 Principles of wave theory

Light propagates itself in a vacuum at a speed of $c_0 = 299{,}792.458$ km/s. To simplify calculations, the value of $300{,}000$ km/s is often assumed as the speed of propagation in air. When propagated through an optically denser medium than air, the propagation speed $c_{medium}$ of light is lower. The ratio of the two propagation speeds is referred to as the refractive index $n$ that is specific to each material. The refractive index value varies according to the wavelength. If the refractive index $n$ of a material is known, the propagation speed $c_1$ is calculated as follows:

$$c_1 = \frac{c_0}{n} \tag{1.1}$$

9

*Example*

In the plastic optical fiber (POF) the refractive index $n_K = 1.49$ (poly-methylmethacrylate), resulting in the following formula for the speed $c_{OF}$

$$c_{OF} = \frac{c_0}{n} = \frac{300.000 \frac{km}{s}}{1,49}$$

$$c_{OF} \approx 200.000 \frac{km}{s}$$

For example, thus a light signal will travel along an optical fiber with a length of 100 m, in 0.5 ms. It should be remembered that the refractive index $n$ depends on the wavelength.

In wave theory, light is represented as an electromagnetic wave with a wavelength $\lambda$ and a frequency $f$. Data transmission via plastic optical fibers uses light with a wavelength in the 400–700 nm band (visible light). The following relationship exists between the variables of wavelength, frequency and speed of light:

$$c = f \cdot \lambda \tag{1.2}$$

## 1.2.2 Optical attenuation

In passing through an optical fiber of length $L$, the light power $P$ decays exponentially:

$$P_L = P_0 \cdot 10^{-\alpha \frac{L}{10}} \tag{1.3}$$

As the light powers extend across many powers of ten, it is customary to make the transition to a logarithmic notation and to specify the attenuation $A$ in decibels (dB):

$$A = 10 \log \frac{P_0}{P_1} \tag{1.4}$$

where $P_0$ signifies the light power in at the start of the optical fiber and $P_1$ the light power in mW at the end of the optical fiber. For the coefficient of attenuation a (kilometric attenuation) with

$$\alpha = \frac{A}{L} \tag{1.5}$$

this produces the measurement unit dB/km. The intensity relative to 1 mW has the unit dBm, corresponding to the following definition:

$$P = 10 \log \frac{P}{1\,mW} \tag{1.6}$$

where P is the light power in mW.

The following examples should help to illustrate this definition:

$$\begin{aligned}
30 \text{ dBm} &= 1 \text{ W} \\
0 \text{ dBm} &= 1 \text{ mW} \\
-30 \text{ dBm} &= 1 \text{ } \mu\text{W} \\
-60 \text{ dBm} &= 1 \text{ nW} \\
-90 \text{ dBm} &= 1 \text{ pW}
\end{aligned}$$

With the aid of the logarithmic notation, we can represent power relationships as differences and calculate attenuations:

$$A = P_0 - P_1 \tag{1.7}$$

where $A$ is the attenuation in dB, $P_0$ the light power at the beginning and $P_1$ the light power at the end of the optical fiber in dBm.

### 1.2.3 Signal transmission in the optical fiber

The basic principle of transmission is based on total reflection. If a ray of light strikes the interface between an optically denser medium with the refractive index $n_1$ and a less dense medium with the refractive index $n_2$, it is either split or totally reflected, depending on the angle of incidence $a_1$ (Fig. 1.2).

When a ray of light strikes the interface between an optically dense medium and an optically less dense medium it is refracted from the perpendicular. As the light strikes the interface at an ever smaller angle, then at a certain angle of incidence $\alpha_1$ the refracted ray assumes an angle of $\alpha_2 = 90°$ to the normal of incidence, i.e. the refracted ray propagates parallel to the interface of the two media. At an even shallower angle of incidence, the refraction becomes total reflection.

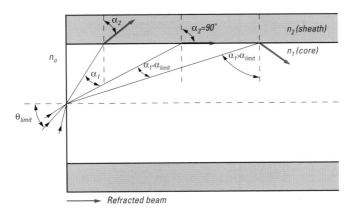

**Fig. 1.2**  Total reflection in a fiber with graded index profile

According to Snell's law of refraction:

$$n_1 \sin\alpha_1 = n_2 \sin\alpha_2 \tag{1.8}$$

Where $\alpha_2 = 90°$ it follows for the critical angle $\alpha_{limit}$ at which the total reflection begins, from formula 1.8:

$$\alpha_{limit} = \arcsin\frac{n_2}{n_1} \tag{1.9}$$

A step index profile optical fiber consists of core glass with the refractive index $n_1$, that is enclosed in cladding glass with the refractive index $n_2$. A condition for a total reflection in the optical fiber is that the cladding with the refractive index $n_2$ is made of an optically less dense medium than the core with the refractive index $n_1$.

The launched light, consisting of a great quantity modes (*eigenwaves*) – which can also be regarded as rays of light with different angles to the axis of the optical fiber – is only capable of propagation if it is launched into the core within a certain angular range. By applying the law of refraction to the end surface and considering the angle ratios according to Fig. 1.2, we obtain the following:

$$n_0 \sin\theta_{limit} = n_1 \sin(90 - \alpha_{limit}) \tag{1.10}$$

Taking into consideration $n_0 = 1$ (air) and formula 1.9, as well as elementary trigonometric formulae, equation 1.10 produces the following:

$$\sin\theta_{limit} = n_1 \cos\alpha_{limit} = n_1 \cos\left(\arcsin\frac{n_2}{n_1}\right) =$$

$$= n_1 \cos\left(\arccos\sqrt{1 - \frac{n_2^2}{n_1^2}}\right) = \sqrt{n_1^2 - n_2^2} \tag{1.11}$$

The sine of the critical angle $\theta_{max}$ is the numerical aperture NA, where

$$NA = \sin\theta_{limit} = \sqrt{n_1^2 - n_2^2} \tag{1.12}$$

Using the definition for the relative refractive index differential $\Delta$

$$\Delta = \frac{n_1^2 - n_2^2}{2n_1^2} \sim \frac{n_1 - n_2}{n_1} \tag{1.13}$$

it is also possible to represent the numerical aperture as follows:

$$NA = n_1 \sqrt{2\Delta} \tag{1.14}$$

The numerical aperture is of major importance when launching light into the optical fiber and when joining optical fibers together. It is determined by the difference between the refractive indices of the core and the cladding.

The value of the input power, however, does not depend only on the numerical aperture of the optical fiber, but also on the core diameter. Ultimately the refractive index profile in the optical fiber core affects the value of the power that can be carried. A change in the refractive index relative to the core diameter means a change in the numerical aperture.

All these variables affect the acceptance area of the optical fiber. This is the range of positions and angles of the optical fiber in which a low-loss propagation of light is possible.

### 1.2.4 Phase space diagram

The possible light ray propagation paths in the optical fiber core are a product of the solution of the wave equation. These solutions of the wave equation are the modes. The number of modes $M$ guided in the optical fiber with the wavelength $\lambda$ can be calculated according to [1.1] as follows:

$$M \approx \frac{1}{2}\left(\frac{2\pi\alpha NA}{\lambda}\right)^2 \tag{1.15}$$

This means, for example, that in a step index POF about 2.4 million modes are capable of propagation. A certain radiance $L\,(A,\,\Omega)$ can be assigned to these modes excited in the optical fiber. This is the power relative to a surface element $A$ and to an element of solid angle $\Omega$. The total power is obtained by integration across over the surface area of the optical fiber core and the associated space angle.

$$P = \int_{A=0}^{A=A_0}\int_{\Omega=0}^{\Omega=\Omega\,(\theta_{\text{limit}})}\!\!\!L(A,\Omega)\mathrm{d}A\,\mathrm{d}\Omega \tag{1.16}$$

The relationships can be explained very clearly with the aid of a phase space diagram (Fig. 1.3).

In this diagram the acceptance area is represented as a function of the standardized surface and the standardized space angle for the step index optical fiber. The integration over the acceptance area gives the power in accordance with equation 1.16.

When coupling optical fibers together, care should be taken to match the acceptance areas, otherwise coupling losses occur. It should also be noted that only one optical source, whose acceptance area lies within the acceptance area of the optical fiber, can be coupled into the optical fiber without loss. This means that ideally the expansion of the source must not exceed that of the optical fiber. At each location the maximum radia-

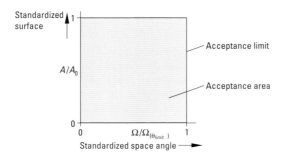

**Fig. 1.3**   Acceptance area in the phase space diagram

tion angle must be equal to or less than the acceptance angle at the associated point of the optical fiber.

Exactly the same behavior is found at the coupling between the optical fiber and the receiver. For this reason, measures to minimize the losses are necessary when transmitting via POFs with a large core diameter (e.g. 980 mm) and a large numerical aperture (see Section 8.3.1).

The coupling between transmitter and optical fiber is therefore critical in optical fibers with a small core diameter and a small numerical aperture.

By means of optical imaging, for example with a lens system with a certain image scale, it is possible to change the shape of the acceptance area (obtaining an extended or compressed rectangle) although the surface size of the acceptance area remains unchanged.

The Helmholtz-Lagrange invariant is responsible for this effect. This states that the product of object size, acceptance angle and refractive index in the object space is equal to the product of the image size, acceptance angle and refractive index in the image space. This means that with a reduced image (image scale less than 1) the image size is smaller, but the acceptance angle is greater. In the case of an enlarged image (image scale greater than 1) the ratios are reversed.

By means of optical imaging with a suitable imaging scale, for example, the radiance distribution of the optical source can be better adapted to the acceptance area of the optical fiber and thus increase the coupling efficiency. Thus an image of a light-emitting diode (LED), which radiates onto the end face in a large angular range with an image scale greater than 1, causes a reduction of the angular range, but the image of the LED is correspondingly enlarged. The principle relationships are illustrated in Fig. 1.4.

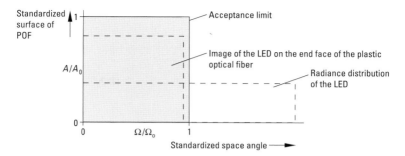

**Fig. 1.4**
Adaptation of the radiance distribution of the LED to the acceptance area of the plastic optical fiber by optical imaging

## 1.2.5 Dispersion and profile

The quality of the optical transmission system is determined not only by the link length that can be bridged, but also by the data rate that can be transmitted.

High data rates demand broadband transmission and reception components (see Section 8), but also broadband optical fibers. The bandwidth in the optical fiber is limited by the dispersion, i.e. by the fact that a pulse launched into the optical fiber disperses as it is propagated in the optical fiber (see Fig. 1.5).

Two pulses emitted with a short interval between them disperse as they propagate along the optical fiber and begin to overlap (see Fig. 1.6).

Beyond a certain point in the transmission the contrast diminishes to such an extent that the individual signals can no longer be separated from one

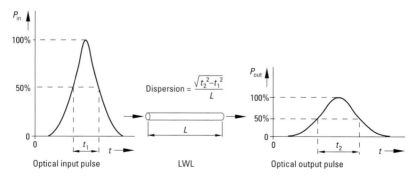

**Fig. 1.5** Pulse dispersion in the optical fiber

15

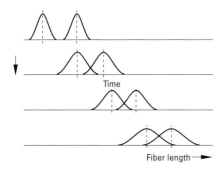

Time

Fiber length ➤

**Fig. 1.6**
Pulse overlapping along the optical fiber

another. As Fig. 1.6 shows, this effect occurs in particular at high bit rates (small time interval between pulses) and over longer transmission lengths.

Dispersion thus causes a broadening of the pulses as they pass along the optical fiber. The pulse broadening limits the bandwidth $B$ and thus the maximum transmission rate (in MHz or Mbit/s) in the optical fiber. As this pulse widening is approximately proportional to the optical fiber length $L$, the bandwidth of the optical fiber diminishes as the length of the optical fiber increases. The following applies by approximation for the bandwidth-distance product:

$$B \cdot L = \text{const} \tag{1.17}$$

*Modal dispersion and profile dispersion*

The most serious problem is the difference in delay of the modes in the optical fiber (modal dispersion). The lowest-order mode that theoretically propagates along the optical axis (see Fig. 1.8) has to cover the distance $L$ in an optical fiber of length $L$. The highest-order mode that can still just be carried, has an angle of inclination $\alpha_{\text{limit}}$ relative to the optical axis (see Section 1.2.3). The path to be covered is then extended to $L/\cos\alpha_{\text{limit}}$. A quick conversion produces the following for the delay difference $\Delta\tau$:

$$\Delta\tau = \frac{L}{2\,c_0\,n_1}\,NA^2 \tag{1.18}$$

It is apparent that the modal dispersion is particularly pronounced where the numerical aperture of the optical fiber is high. This is the case in particular for POFs with typical $NA$ values of $0.47$. A reduction of the mode dispersion in the step index profile optical fiber is only possible by reducing the numerical aperture.

What other possible ways are there of reducing the modal dispersion? One solution is to increase the propagation speed of the rays near the margin, the material at the margin having a lower refractive index. This idea led to the development of the graded index profile optical fiber in which the refractive index varies across the cross section of the core. As the core radius grows the refractive index diminishes, the material becomes optically less dense, the propagation speed increases. In other words, the greater the distance from the optical axis, the faster the signal propagates. In this way the time differences $\Delta\tau$ between the low and high order modes can be minimized.

The core refractive index as a function of the radius $r$ can be generally represented as a power law profile:

$$n(r)^2 = n_1^2 \left(1 - 2\Delta \left(\frac{r}{a}\right)^g\right), \quad 0 \le r \le a$$
$$n(r)^2 = n_2^2 \qquad\qquad , \quad r \ge a \qquad\qquad (1.19)$$

where $a$ is the core radius and $g$ is the profile exponent. With $g = \infty$ the graded index profile is also included. For all profiles, $n(r = 0) = n_1$ and $n(r = a) = n_2$ (substituting $\Delta$ according to equation 1.13). Equation 1.19 is illustrated on Fig. 1.7.

It is now possible to find an optimum profile exponent $g_{opt}$ (see formula 1.20) [1.2], in which the modal dispersion is minimal and thus the delays of all modes are approximately the same.

$$g_{opt} = 2 - 2\Delta \sim 2 \qquad\qquad (1.20)$$

This results in the widely used graded index profile optical fiber. The modes travel through the optical fiber with a sinusoidal motion and the delay differences are balanced out. With such a profile, the dispersion can be reduced in theory by $10^3$ and in practice by $10^2$ in comparison with step index profile optical fibers. This discrepancy between theory

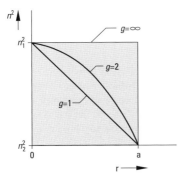

**Fig. 1.7**   Power-law profile

and practice arises from the fact that the dispersion depends largely on profile changes. From a manufacturing viewpoint, certain deviations from an ideal profile must always be taken into consideration.

It should be noted that the optimum profile exponent $g_{opt}$ depends on the refractive index difference $\Delta$, but that $\Delta$ is a function of core and cladding refractive indices which in turn depend on the wavelength and the doping level. The physically unavoidable remainder of the modal dispersion in the graded index profile optical fiber is referred to as profile dispersion.

There is only one way to avoid profile dispersion: an optical fiber must be used that carries only one mode – the single-mode fiber. To achieve this, the acceptance area in the optical fiber must be severely limited so that the wave equation only delivers a single solution (mode). The typical core radius of such a fiber is 9 μm and the typical numerical aperture is 0.1. At these dimensions it is necessary to switch from the geometrical-optical representation to the wave-optical representation. The propagation in the optical fiber now has to be presented as a pulse with Gaussian distribution.

This Gaussian distribution has its maximum on the fiber axis and declines almost exponentially as the radius increases. It is not cut off by the core-cladding interface, but extends far into the cladding. For this reason, it is not sufficient to characterize the single-mode fiber by using the core radius. The determining factor is the mode field radius $w_0$ that is defined by a drop in the power to $1/e^2$. The relationship between core radius and mode field radius in the single-mode fiber depends on the profile exponent.

Equally, the "numerical aperture" of the optical fiber is obtained from the sine of the angle under which the far-field intensity has dropped to $1/e^2$ (the quotation marks are used because it does not refer to an angle of a beam in the conventional definition of the numerical aperture).

Schematic representations of the three basic types of optical fiber are shown in Fig. 1.8.

*Material dispersion and waveguide dispersion*

Even after the introduction of the single-mode fiber it was noticed that there were limitations to the transmittable bandwidth as before. These effects are small in comparison with the modal dispersion and only occur if this is suppressed. It concerns the material dispersion and waveguide dispersion which, like the profile dispersion, combine to form the chromatic dispersion.

The material dispersion arises from the fact that the group refractive index $n_{gr}$ is wavelength dependent and, as shown in Equation 1.1, determines the group speed. The group speed is the determining variable for

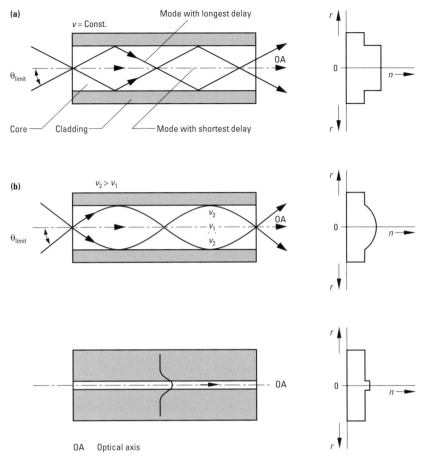

**Fig. 1.8**
Basic types of optical fiber: (a) Fiber with step index profile; (b) Fiber with parabolic index profile; (c) Single-mode fiber

the propagation of the light wave in the optical fiber, in contrast to the phase speed which describes the propagation speed of the wave surface (phase front). The group refractive index is calculated from:

$$n_{gr} = n - \lambda \frac{dn}{d\lambda} \tag{1.21}$$

Depending on the spectral width $\Delta\lambda$ of the transmitter, this results in a corresponding broadening of the pulse (see Fig. 1.9).

19

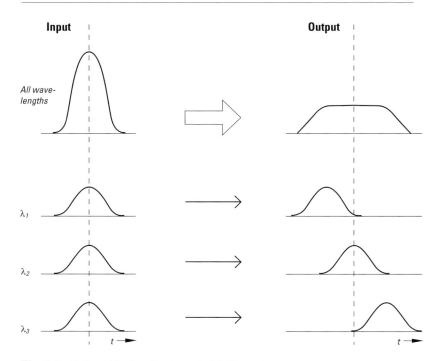

**Fig. 1.9**  Pulse widening due to material dispersion

A further type of dispersion, waveguide dispersion, is caused by the wavelength dependence of the light distribution of the basic mode on the core and cladding. The greater the wavelength, the more the wave spreads out and the further it penetrates into the cladding. A further portion of the light is guided in the cladding and as a result of the low refractive index has a higher propagation speed. This produces a balanced mean value from the propagation speeds of the core and cladding. This is dependent on the wavelength. An increase in the bandwidth in the transmission with single-mode fiber is possible by reducing the spectral width of the transmitter (use of lasers with only one very small mode) and/or by compensating for the material dispersion with the waveguide dispersion.

*Summary of dispersion characteristics*

The most frequent types of dispersion, dependent on the transmission system, are listed in Table 1.1. The specified numerical values for the bandwidth-distance product are rough guide values that depend on a host of system parameters.

**Table 1.1**  Dispersion in various glass optical fiber systems

|  | Multimode optical fiber | | Single mode optical fiber |
|  | Step index profile | Parabolic index profile | |
| --- | --- | --- | --- |
| LED | Mode dispersion $(\tau_{mod} \gg \tau_{chrom})$ | Chromatic dispersion | Chromatic dispersion $(\tau_{chrom} \gg \tau_{mod})$ |
|  | $B \cdot L \approx 50$ MHz $\cdot$ km | $B \cdot L \approx 500$ MHz $\cdot$ km | $B \cdot L \approx 1$ GHz $\cdot$ km |
| Laser diode | Mode dispersion $(\tau_{mod} \gg \tau_{chrom})$ | Profile and chromatic dispersion | Chromatic dispersion $(\tau_{chrom} \gg \tau_{mod})$ |
|  | $B \cdot L \approx 50$ MHz $\cdot$ km | $B \cdot L \approx 1$ GHz $\cdot$ km | $B \cdot L \approx 100$ GHz $\cdot$ km |

When several dispersion effects are overlaid, the total dispersion $\tau_{total}$ is calculated from the individual dispersion components of the chromatic dispersion $\tau_{chrom}$ and the mode dispersion $\tau_{mod}$ as follows:

$$\tau_{total} = \sqrt{\tau_{chrom}^2 + \tau_{mod}^2} \qquad (1.22)$$

The mode dispersion depends greatly on the numerical aperture of the input. It diminishes with the square of the numerical aperture. With launch apertures of $\sim 0.1$ and large spectral widths of the source, the chromatic dispersion is no longer negligible in comparison with the mode dispersion.

## 1.3 Physical values for transmitters and receivers

The transmit power $P$ is specified either in W or derived values such as mW and μW or in dBm according to formula 1.6.

When converting the light power at the end of the optical transmission route into an electrical signal, receiver sensitivity is the determining variable. This specifies which light power (in Watts) affects which current (in Amps). Typical receiver sensitivities are in the range of $0.5-1$ A/W. In the case of receivers that supply a prepared signal, e.g. TTL, at the output, the receiver sensitivity is given in dBm. In these cases it is particularly easy for the user to perform budget examinations (see Chapter 13).

Transmitters and receivers for optical fibers are considered in detail in Chapter 8.

# 2 Overview of the most common optical fibers

The different types of optical fiber are categorized either by their construction, e.g. step index profile or graded index profile, or by a special property, e.g. single-mode or multimode. The two methods of designation are used side by side with equal status by specialists in the field. Table 2.1 gives an overview of the physical properties of the individual fibers, with particular attention being paid to the common types of fiber used today, the plastic optical fiber (POF), the polymer clad fibers (PCFs) and the glass optical fibers (GOF).

**Table 2.1**  Overview of the commonest optical fibers

|  | POF | PCF-OF |  | Glass OF |  |
|---|---|---|---|---|---|
| Fiber type | Multimode step index | Multimode step index | Multimode graded index | Multimode graded index | Single-mode step index |
| Fiber core material | Plastic | Glass | Glass | Glass | Glass |
| Fiber cladding material | Plastic | Plastic | Glass | Glass | Glass |
| Core/cladding diameter in µm | 980/1000 | 200/230 | 62,5/125 | 50/125 | 9/125 |
| Diameter of first buffer tube in µm | 2200 | 500 | 250 | 250 | 250 |
| Numerical aperture | 0.47 | 0.36 | 0.27 | 0.2 | approx. 0.5 |
| attenuation coefficient $\alpha$ in dB/km | | | | | |
| at 660 nm | 230 | 7 | – | – | – |
| at 850 nm | 2000 | 6 | ≤3.5 | ≤3.0 | – |
| at 1300 nm | – | – | ≤0.80 | ≤0.70 | ≤0.40 |
| typical wavelength in nm | 660 | 660/850 | 850/1300 | 850/1300 | 1300 |
| Bandwidth-distance product in MHz · km | | | | | |
| at 660 nm | 1 | – | – | – | – |
| at 850 nm | – | ≥17 | ≥200 | ≥400 | – |
| at 1300 nm | – | – | ≥600 | ≥600 | >10000 |
| Chromatic dispersion at 1300 nm | | | | | ≤3,5 ps/km · nm |

**Table 2.2**   Application examples of the individual optical fiber

| Fiber core diameter in μm | Applications | Typical distance | Typical data rate |
|---|---|---|---|
| 10 | Telecommunications | over 10 km | Mbit/s/Gbit/s |
| 50 and 62.5 | LANs in buildings | up to 4 km | < 155 Mbit/s |
| 200 | Industrial LANs | up to 2 km | < 100 Mbit/s |
| 980 | Industrial LANs, vehicles, LANs in buildings | up to 100 m | < 40 Mbit/s |

The optical attenuation has a crucial bearing on the length of the transmission link that can be implemented. The bandwidth-distance product provides information about the transmittable data rate. These transmission characteristics thus determine the possible field of application of the optical fiber. Typically bridgeable distances and data rates, as well as applications are listed in Table 2.2.

## 2.1 Comparison of the properties of different transmission media

Why are optical fibers of interest as transmission media in the first place? This chapter will attempt to answer this question.

Table 2.3 compares some key properties of plastic optical fiber, glass optical fiber and copper conductors. The first group deals with electromagnetic compatibility, electrical isolation, immunity to eavesdropping and the risk of explosion in hazardous areas. In all these respects, glass and plastic optical fiber have the same advantages, arising from the fact that the photons as carriers of the information in the optical fiber have no electrical charge like the electrons which carry the information in copper conductors.

The second group of characteristics comprises the external and mechanical properties. A small bending radius and high flexibility are the advantages that make plastic optical fiber an attractive alternative to glass optical fiber. The low weight of optical fibers compared with copper conductors is an advantage in almost every application.

Compared to glass optical fiber, the connectorizing of plastic optical fibers is considerably easier due to their greater fiber cross section. The advantages of glass in terms of bandwidth are unsurpassed. The bandwidth of plastic optical fiber is, however, usually adequate over shorter distances, e.g. in industrial and domestic applications or in vehicles.

**Table 2.3**
Comparison of properties of plastic and glass optical fibers and copper conductors

|  | Plastic optical fiber | Glass optical fiber | Copper conductor |
|---|---|---|---|
| Electromagnetic compatibility (EMC) | ++ | ++ | − |
| Electrical isolation | ++ | ++ | − |
| Immunity to eavesdropping | + | + | − |
| Risk in hazardous environments | ++ | ++ | − |
| Low weight | + | + | − |
| Flexibility | + | − | + |
| Small bending radius | + | − | + |
| Connectorization | ++ | − | + |
| Bandwidth | + | ++ | + |
| Optical attenuation | − | + | |
| Cost | ++ | − | + |

++ very good; + good; − unsatisfactory

## 2.2 The advantages of plastic optical fiber

• Large fiber cross-section

Due to the large fiber cross-section the positioning of the plastic optical fiber at the transmitter or receiver presents no great technical problems. In contrast to glass optical fiber, whose cross-sections are in the order of μm, no expensive precision components are required for centering the plastic optical fibers.

• Relative immunity to dust

In industrial environments in particular, where dust is part of everyday life, the large fiber diameter proves to be an advantage. Even when the fibers are properly handled, dust can get onto the fiber end face, affecting the input and output optical power in every case. But with plastic optical fibers, minor contamination does not necessarily result in failure of the transmission route. For this reason, the plastic optical fibers can readily be connectorized on site in an industrial environment.

• Simple use

The 1 mm thick plastic optical fiber is easier to handle than, for example, a 62.5/125 μm thick glass fiber. Plastic optical fibers are considerably less problematical to handle than glass fibers. In particular, glass fibers tend to break when bent around a small radius, which is not the case with plastic optical fibers. In simple terms, "glass is more brittle than plastic".

• Advantages of polymethylmetacrylate (PMMA) fiber core material

PMMA is easy to cut, grind and melt. Little time is required therefore for processing the end faces to achieve a clean, smooth and grooveless surface. In addition, despite its relatively large cross-section, plastic optical fiber has an extraordinarily high flex resistance. This facilitates the cost-effective use of plastic optical fiber even under severe loading conditions that are often encountered in mechanical engineering applications.

• Low cost

All the stated properties of plastic optical fiber ultimately mean that the components for connection to transmitters and receivers (connectors, housings) are relatively economical. The uncomplicated processing of the end faces can be performed extremely cost effectively, especially after assembling in the field.

# 3 Construction and manufacture of plastic optical fiber

## 3.1 Plastic optical fiber materials

### Materials for the fiber core

The most commonly used material today for fiber cores is polymethyl-metacrylate (PMMA). In addition, other materials such as polystyrene (PS) and polycarbonate (PC) are used as fiber core materials for specialized applications. The most important properties of these materials are listed in Table 3.1 and their chemical structure is represented in Fig. 3.1.

PS is very brittle and therefore, despite its very low optical attenuation compared with other materials, seldom used. Due to its high resistance to temperature, PC is particularly suitable for applications subject to high ambient temperatures (e. g. in vehicles), but it has a significantly higher optical attenuation than PMMA. Compared with PMMA, however, PC has a significantly higher alternate flex resistance at small bending radii [3.1].

**Table 3.1** Properties of fiber core materials

| Material | Refractive index $n_1$ | Optical attenuation/wavelength | Glass transition temperature |
|---|---|---|---|
| PMMA | 1.49 | 70–100  dB/km / 570 nm<br>125–150 dB/km / 650 nm | 105 °C |
| PC | 1.58 | 700 dB/km / 580 nm<br>600 dB/km / 765 nm | 150 °C |
| PS | 1.59 | 90  dB/km / 580 nm<br>70  dB/km / 670 nm | 100 °C |

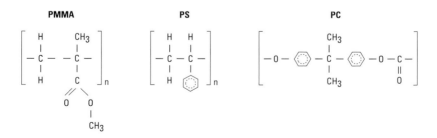

**Fig. 3.1**   Chemical structure of PMMA, PS and PC

The search for suitable new materials for plastic optical fibers is concentrated on the one hand on minimizing the optical attenuation and, on the other hand, on materials that are resistant to high temperatures as well as offering a low optical attenuation.

Laboratory experiments with PMMA, in which hydrogen atoms are replaced by fluorine or deuterium atoms, indicate a significantly reduced optical attenuation of the fibers. Theoretically an attenuation of 9.1 dB/km at 680 nm is achievable by using deuterium [3.2]. This has been investigated in greater detail as part of the development work on the manufacture of the plastic graded index fibers. It has thus been possible to measure an optical attenuation of less than 50 dB/km in the 500–1300 nm waveband on a fluorized plastic graded index fiber [3.3].

### Cladding materials

Fluropolymers are now used for cladding. From a physical viewpoint, when selecting these materials, it is first important that the refractive index $n_2$ of the cladding is less than the core refractive index $n_1$. In addition, the difference in refractive index has a crucial bearing on the size of the numerical aperture (see Equation 1.12). For modern PMMA fibers, fluoropolymers with a refractive index in the order of 1.35–1.42 are used. PMMA is sometimes used as the cladding for polystyrene.

## 3.2 Types of fiber

### Step index plastic optical fiber

Only step index plastic optical fibers are commercially available at present. Fibers of this type are specified in IEC 60793–2 (see Table 3.2)

A category A4d is currently being discussed, which differs from category A4a in respect of the numerical aperture ($NA = 0.33 \pm 0.03$) – resulting in a higher bandwidth-distance product ($\geq 100$ MHz $\cdot$ 100 m) – and of the optical attenuation ($\leq 200$ dB/km). Due to the lower numerical aperture, this fiber is often referred to as "low NA" fiber. The lower NA of this

**Table 3.2**  Standardized plastic optical fiber types

| Category | A4a | A4b | A4c |
|---|---|---|---|
| Core diameter | typically 10 to 20 µm less than cladding diameter | | |
| Cladding diameter in µm | $1000 \pm 60$ | $750 \pm 45$ | $500 \pm 30$ |
| Numerical aperture | $0.5 \pm 0.15$ | | |
| Attenuation in dB over 100 m | $\leq 40$ | | |
| Bandwidth-distance product in MHz $\cdot$ 100 m | $\geq 10$ | | |

Fiber core (PMMA)
Fiber cladding

**Fig. 3.2** Construction of a plastic optical fiber with step index profile

fiber however causes the bending radius to have a greater effect on the attenuation than with the A4a fiber. This must be taken into account when using such fibers.

Apart from the standardized types of fiber, other step index profile POFs with diameters of 75 μm, 125 μm, 380 μm, 1500 μm, 2000 μm and 3000 μm are commercially available.

The construction principle of a step index plastic optical fiber is shown in Fig. 3.2.

The fiber core is used for guiding the light waves. As the cladding immediately adjacent to the core has a lower refractive index than the core, it permits total reflection and thus causing the light to be guided in the core. Any damage to the cladding therefore causes output coupling loss, resulting in increased optical attenuation.

The fiber code is explained below, consistent with DIN VDE 0888, Part 4, using the example of a 1000 μm fiber:

*Code*

F – P980/1000 150A10
F – Fiber
P – Plastic optical fiber with step index profile
980/1000 – Core/cladding diameter
150 – attenuation coefficient in dB/km
A – Wavelength 650 nm
10 – Bandwidth-distance product 10 MHz · 100 m

*Technical characteristics*

Currently, plastic optical fibers with a PMMA core are available with the following typical characteristics:

Core diameter: 980 mm
Cladding diameter: 1000 mm
Tensile strength: 5 N
Minimum bending radius: 20 mm
Optical attenuation: 130–150 dB/km
Numerical aperture: 0.5
Acceptance angle: 30°
Bandwidth-distance product: 40 MHz · 100 m

In the case of attenuation coefficients, it should be noted that their measurements are performed at 650 nm using a mode mixer.

**Graded index profile plastic optical fiber**

Graded index profile plastic optical fibers are currently under development and the possible fiber diameters are discussed in [3, 4] which states that, for systems with a data transmission rate of up to 1 Gbit/s, fiber diameters of between 600 mm and 1200 mm are suitable for graded index plastic optical fibers. Further investigations have reported on successful experiments to transmit data at 2.5 Gbit/s over 100 m and estimate the bandwidth of graded index plastic optical fibers to be about 2 GHz over a distance of 1 km [3.5; 3.6].

If the mass production of such fibers one day becomes possible, then plastic optical fibers can be used in new applications such as local area network (LAN) cabling.

A summary of the current status of development can be found in [3.7].

**Single-mode plastic optical fiber**

The manufacture of single-mode POF has also been laboratory tested. An optical attenuation of 200 dB/km has been achieved at a wavelength of 652 nm [3.8]. A further development in this field does not appear practical, as it is precisely the large cross section that is the advantage of plastic optical fiber.

# 3.3 Manufacturing procedures

Four basic steps must be performed in the manufacture of plastic optical fibers:

▷ cleaning of the starting materials,

▷ polymerization,

▷ forming of the fiber geometry and

▷ application of the fiber cladding.

The cleaning of the monomer starting material is of crucial importance to the optical attenuation quality of the fibers. The most common causes of contamination in the starting materials are

▷ inhibitors that are added to the material to prevent premature polymerization,

▷ by-products from the manufacture of monomers,

▷ water, metal and dust particles.

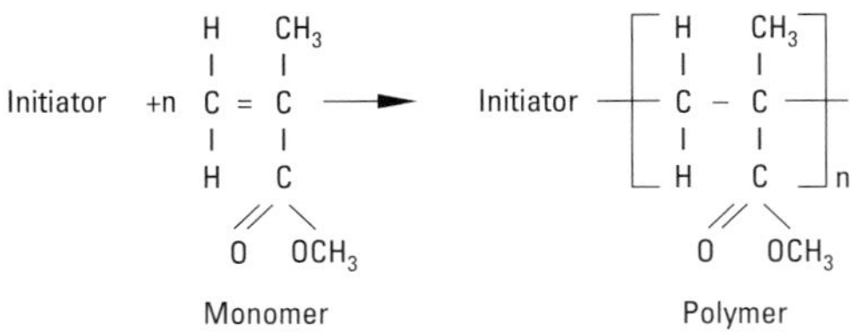

**Fig. 3.3**  Polymerization of PMMA

During polymerization, chain-like macro-molecules (= polymers) are created from numerous individual molecules (= monomers). This process requires additives such as initiators and polymerization regulators. Due to the stringent purity demands on the polymer, it is important to select a procedure that requires the minimum possible amounts of these additives. For the same reason, contamination from the apparatus must also be kept to a minimum during this process.

The basic sequence of the chemical reaction is shown in Fig. 3.3, using the example of PMMA, where $n$ specifies the degree of polymerization, i.e. the number of cross-linked monomers.

The following procedures are familiar in the manufacture of plastic optical fibers:

▷ Fiber drawing from the preform,

▷ Batch extrusion,

▷ Continuous extrusion and

▷ Melt-spinning process.

**Fiber drawing from the preform**

This procedure is already familiar from the manufacture of glass optical fibers. First a preform is manufactured, e.g. using the batch extrusion method (see Fig. 3.4). The preform consists of a polymer cylinder that is concentrically sheathed in the cladding. This preform is heated until the fiber can be drawn from it. The overall process is operated in batch mode and is relatively expensive. Graded index fibers can be manufactured more simply by other methods. In [3.9] this procedure is used, for example, to manufacture plastic optical fibers with a polystyrene core. This procedure is very well suited to the production of graded index plastic optical fibers, as the refractive index profile can be manufactured more easily than with the extrusion process. It also makes use of the experience gained in the manufacture of glass fibers. Successful experiments in this area are described, for example, in [3.7].

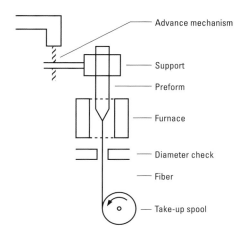

— Advance mechanism

— Support

— Preform

— Furnace

— Diameter check

— Fiber

— Take-up spool

**Fig. 3.4**
Arrangement for the manufacture of plastic optical fibers by drawing from a preform

## Batch extrusion

In a vacuum the monomer is first introduced to the polymerization container (see Fig. 3.5) by means of distillation. The initiator and polymerization regulator are then added in the same way. The polymerization takes place in the container at about 180 °C. Once this has been completed, the container is sealed and the molten polymer is extruded through a nozzle by injecting nitrogen under pressure. As the fiber leaves the nozzle the cladding is immediately applied. The procedure is not used on a large scale because of its batch operation.

## Continuous extrusion

The arrangement in Fig. 3.6 is suitable for manufacturing plastic optical fiber continuously on a large scale. In a heated reactor vessel, the mixture of polymer, initiator and polymerization regulator are 80% pre-polymerized. This mixture is then pumped into an extruder where the gas is extracted, residual monomer is drawn off and returned to the reactor for pre-polymerization. At the output of the extruder the polymer is forced through a nozzle which gives the fiber its geometric form. A second extruder is used for cladding the fiber.

## Melt-spinning process

In this procedure the polymer is melted and then pressed through a spinneret (see Fig. 3.7). Some of the holes in the spinneret are used for the geometrical formation of the fibers, the others as a feed for the cladding

31

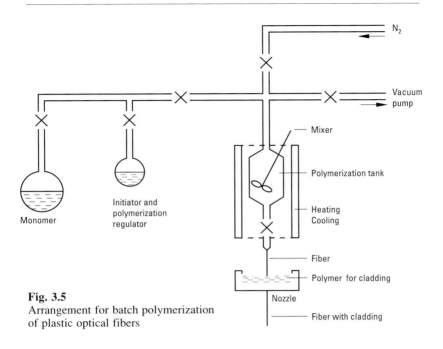

**Fig. 3.5**
Arrangement for batch polymerization
of plastic optical fibers

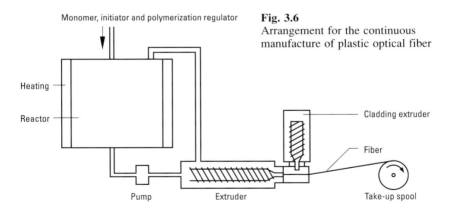

**Fig. 3.6**
Arrangement for the continuous
manufacture of plastic optical fiber

polymer. In this way a fully clad fiber is obtained at the output of the spin-neret. By using a spinneret with several holes, it is possible to manufacture several fibers simultaneously. This efficient process facilitates extremely high drawing speeds, but such a system is very expensive to set up.

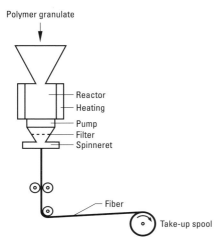

Polymer granulate

Reactor
Heating
Pump
Filter
Spinneret

Fiber

Take-up spool

**Fig. 3.7**
Arrangement for the manufacture
of plastic optical fiber using the
melt-spinning process

In all the procedures described above, the fiber is subjected to a stretching process after geometric formation, in which the polymer molecules are given a special orientation. This process gives the fiber the required diameter and has a decisive influence on the mechanical properties of the fiber, e. g. the tensile strength.

# 4 Fiber transmission properties and measurement procedures

The attenuation and the bandwidth-distance product are the most important parameters for identifying the transmission properties of an optical fiber.

The measurement procedures for plastic optical fiber have, for the most part, been based on the procedures that apply to glass optical fibers, with some adaptation of the test conditions. In Europe these procedures are described in EN 187000/A1 or EN 188000 and in IEC 60793-1 and in Japan in JIS C 6863.

## 4.1 Optical attenuation

### 4.1.1 Causes

During propagation in the waveguide, light loses power in function to its wavelength $\lambda$. It experiences attenuation $A$ $(\lambda)$ which is measured in dB (see Formula 1.4). The lower the attenuation, the greater the measurable light signal at the end of the optical fiber. The attenuation therefore has a crucial effect on the distance that can be spanned with an optical fiber. The attenuation of an optical waveguide is usually specified by the attenuation coefficient $\alpha$ $(\lambda)$ in dB/km.

The attenuation is caused mainly by the following effects:

▷ Scattering $(\alpha_S)$

▷ Absorption $(\alpha_A)$ and

▷ Radiation loss $(\alpha_L)$

The total of the individual components produces the attenuation coefficient $\alpha$, where

$$\alpha = \alpha_S + \alpha_A + \alpha_V \tag{4.1}$$

Scattering is caused on the one hand by inhomogeneities (fluctuations of density, concentration) in the optical fiber material (intrinsic, i.e. unavoidable, losses) and, on the other hand, by extrinsic (avoidable) losses such as inclusions and contamination. As the PMMA used for plastic optical fibers is extremely pure, the intrinsic losses dominate.

The losses caused by scattering $a_S$ can be described with good approximation by the Rayleigh scattering law:

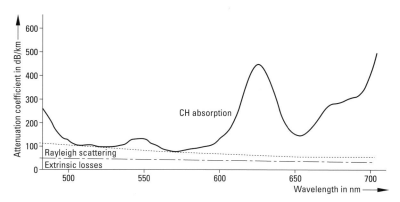

**Fig. 4.1**  Graph of the attenuation coefficient $\alpha$ ($\lambda$) of a PMMA fiber

$$\alpha_S \sim \frac{1}{\lambda^4} \tag{4.2}$$

Whereas the scattering takes place at all wavelengths, the absorption only occurs at certain wavelengths. In the case of absorption, the photons interact with electrons and molecules of the fiber material, causing the photon energy, i.e. the light energy, to be absorbed. The absorption by electrons is negligible in PMMA, whereas in CH compounds it has a crucial effect on the optical attenuation (Fig. 4.1) Extrinsic absorption losses occur as a result of chemical impurities. Radiation losses on the other hand arise from bending of the optical fiber, fluctuations in the fiber diameter and defects in the core-cladding interface and are therefore of an extrinsic nature.

### 4.1.2  Measurement of optical attenuation

In practice, the individual components of attenuation are of less interest than the total losses that cause optical attenuation.

In the transmitted light procedure, an optical fiber of length $L$ in m is connected to a light source with a defined wavelength of power $P_0$ in dBm. At the end of the optical fiber the light power is measured in dBm. The power loss, i.e. the optical attenuation $A$ in dB can be determined from the difference between $P_0$ and $P_L$ as

$$A = P_0 - P_L \tag{4.3}$$

The attenuation coefficient $\alpha$ in dB/m is equal to the attenuation $A$ over a length $L$ of 1 m:

$$\alpha = \frac{A}{L} = \frac{P_0 - P_L}{L} \tag{4.4}$$

35

To determine the light intensity at the beginning and end of the optical fiber, the insertion loss technique and the cut-back technique are usually chosen.

In the *insertion loss technique* the light power $P_0$ in dBm is measured at the end of a short section of optical fiber. The optical fiber section should match the optical fiber under test in both construction and properties. The optical fiber to be tested is then connected and the light power $P_L$ in dBm is determined at the end of the fiber. In calculating the attenuation coefficient $a$ in dBm, it is then only necessary to consider the attenuation $A_{ref}$ of the short section of fiber $L_{ref}$ which acts as a reference, as follows:

$$\alpha = \frac{P_0 - P_L - A_{ref}}{L} \qquad (4.5)$$

When performing this measurement, special attention should be paid to equal excitation conditions and clean connectors, as these have a decisive effect on the repeatability and precision of the procedure.

An improvement can be achieved by using the *cut-back method,* in which an optical fiber of length $L$ is connected to the light source and receiver and the light power $P_L$ is measured. The optical fiber is then detached from the receiver and cut back 1 m beyond the light source (corresponding to $L_{ref}$). This reference section is reconnected to the receiver and the value $P_0$ is measured. Thus the attenuation coefficient $\alpha$ in dB/m is calculated as follows:

$$\alpha = \frac{(P_0 - P_L)}{(L - L_{ref})} \qquad (4.6)$$

When specifying the attenuation and attenuation coefficient, the wavelength at which these values were measured must be stated.

Another method is the *backscattering technique,* in which the light is launched into one end of the optical fiber and also received at the same end. The received light is created in the same optical fiber by Rayleigh scattering. This method thus enables localized defects to be detected and pinpointed. The amount of backscattered light, however, is very small in comparison with the input light, as it is attenuated again on its return path. Due to the high attenuation of the plastic optical fiber, this method requires a very powerful transmitter and correspondingly sensitive receiver. Such equipment is not commercially available at present and therefore this method is not considered for measurements on plastic optical fiber. On the other hand, it is the most widely used method for determining the attenuation in glass optical fibers.

### 4.1.3 Filter effect

The filter effect occurs specifically in plastic optical fibers and its result is that the attenuation coefficient can be length-dependent, depending on the transmitter used.

How does this effect occur? The term *filter effect* itself suggests a phenomenon in which certain parts of the light are "filtered", so we will first consider the attenuation spectrum of a plastic optical fiber.

As can be seen from Fig. 4.2, the optical attenuation depends to a great extent on the wavelength.

Three areas can be found in which data transmission is practical due to the local attenuation minima. As with glass optical fibers, we call these areas "optical windows" and they are shown in Table 4.1 in comparison with those of the glass optical fiber.

In the third and typically used optical window of 650–670 nm a rise in attenuation can be observed from 130 dB/km at 650 nm to 380 dB/km at 670 nm. Such a sudden change cannot be observed at wavelengths around 570 nm and 520 nm. Due to the low optical attenuation, therefore, these areas would appear suitable for data transmission. However, as such transmitters are only available so far for a few specialized applications, this area is only rarely used at present.

Let us look then at the emission spectrum of a light emitting diode LED (see Fig. 4.2), that is used in practice as a transmitter. The typical *full width at half maximum* (FWHM) of such an LED is 20–30 nm. The peak wavelength is in the range of 650–670 nm.

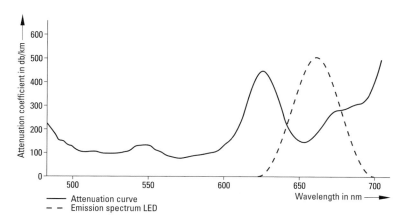

**Fig. 4.2**
Attenuation spectrum of a plastic optical fiber and emission spectrum of an LED; FWHM (full width at half maximum) = 30 nm

**Table 4.1**
Optical windows for transmission for plastic optical fiber of PMMA and for glass optical fiber at different wavelengths

|  | Optical window | Wavelength minimum in nm | Typical minimum attenuation coefficient in dB/km* |
|---|---|---|---|
| Plastic optical fiber | 1 | 520 | 73 |
|  | 2 | 570 | 66 |
|  | 3 | 650 | 130 |
| Glass optical fiber | 1 | 850 | 4 |
|  | 2 | 1300 | 0,4 |
|  | 3 | 1550 | 0,2 |

\* The specified values are typical and depend on the fiber quality and measurement conditions.

The area under the bell-shaped curve corresponds to the entire optical transmit power.

The effect of the close relationship of the attenuation coefficient (see Fig. 4.2) to the wavelength is that the individual components of the emitted light power are attenuated to varying degrees. For example, the light components at 650 nm are attenuated less than all other components above or below 650 nm. If one then considers the spectral distribution at the end of a plastic optical fiber (see Fig. 4.3) the peak wavelength $\lambda_{peak}$ is shifted toward 650 nm, the minimum attenuation, and the full width at half maximum becomes smaller.

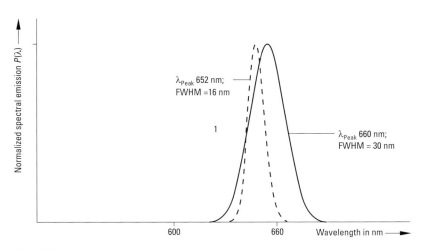

**Fig. 4.3**
Normalized spectral distribution of the transmit power at the beginning and at the end of a 100 m long plastic optical fiber (calculated)

The calculation shows that at the end of a 100 m plastic optical fiber, the peak wavelength of 660 nm is shifted toward 652 nm and the full width at half maximum is halved (see Fig. 4.3).

The characteristic of the attenuation relative to the length of the plastic optical fiber can be calculated as described below.

The starting point of our considerations is Formula 1.4, remembering that the value $P$ depends on the wavelength.

$$A(L) = 10 \log \frac{P_0(\lambda)}{P_1(\lambda)} \tag{4.7}$$

For the calculation of the power launched into the OWG $P_0(\lambda)$, whose spectral distribution can be described approximately by a Gaussian distribution, we specify the peak wavelength $\lambda_{peak}$ and the full width at half maximum $\Delta\lambda_{FWHM}$. Coupling losses are not considered at this stage:

$$P_0(\lambda) = e^{\frac{-4 \ln 2}{\Delta\lambda_{FWHM}^2} \cdot (\lambda - \lambda_{peak})^2} \tag{4.8}$$

The attenuation coefficient $\alpha(\lambda)$ is available as a measured data series (measured monochromatically; see Table 4.2).

By transposing formula 4.7 we obtain $P_L$, $P_0$ being used as described above.

$$P_L(\lambda) = P_0 \cdot 10^{-\frac{\alpha(\lambda)L}{10}} \tag{4.9}$$

Thus the attenuation $A$ of an optical fiber of length $L_1$ is calculated, integrating via $\lambda$, as follows:

$$A_{L_1}(\lambda) = 10 \log \frac{\int_0^\infty P_0(\lambda)\, d\lambda}{\int_0^\infty P_{L_1}(\lambda)\, d\lambda} \tag{4.10}$$

Where $P_0$ is calculated using formula 4.8 and $P_1$ with formula 4.9.

The result of this calculation demonstrates that the attenuation coefficient (see Fig. 4.4) is a distance-dependent variable when using an LED as a transmitter. As the distance increases this dependence diminishes.

The values presented in Table 4.3 can be derived from the same calculation for various peak wavelengths for 50 m plastic optical fibers.

Logically, such a dependence does not exist when using a monochromatic source as a transmitter.

**Table 4.2**
Typical spectral progression of the attenuation coefficient of a plastic optical fiber

| Wavelength $\lambda$ in nm | Attenuation coefficient $\alpha$ in dB/km | Wavelength $\lambda$ in nm | Attenuation coefficient $\alpha$ in dB/km | Wavelength $\lambda$ in nm | Attenuation coefficient $\alpha$ in dB/km |
|---|---|---|---|---|---|
| 400 | 162 | 554 | 94 | 628 | 402 |
| 410 | 148 | 556 | 88 | 630 | 361 |
| 420 | 134 | 558 | 78 | 632 | 326 |
| 430 | 119 | 560 | 73 | 634 | 285 |
| 440 | 109 | 562 | 71 | 636 | 244 |
| 450 | 101 | 564 | 70 | 638 | 209 |
| 460 | 92 | 566 | 69 | 640 | 184 |
| 470 | 88 | 568 | 69 | 642 | 166 |
| 480 | 90 | 570 | 69 | 644 | 147 |
| 490 | 86 | 572 | 71 | 646 | 138 |
| 500 | 76 | 574 | 73 | 648 | 134 |
| 502 | 75 | 576 | 73 | 650 | 132 |
| 504 | 74 | 578 | 76 | 652 | 140 |
| 506 | 75 | 580 | 79 | 654 | 150 |
| 508 | 74 | 582 | 78 | 656 | 166 |
| 510 | 73 | 584 | 82 | 658 | 181 |
| 512 | 73 | 586 | 84 | 660 | 199 |
| 514 | 73 | 588 | 84 | 662 | 217 |
| 516 | 73 | 590 | 88 | 664 | 234 |
| 518 | 73 | 592 | 94 | 666 | 250 |
| 520 | 73 | 594 | 103 | 668 | 262 |
| 522 | 73 | 596 | 111 | 670 | 272 |
| 524 | 75 | 598 | 123 | 672 | 277 |
| 526 | 78 | 600 | 136 | 674 | 285 |
| 528 | 79 | 602 | 150 | 676 | 287 |
| 530 | 82 | 604 | 168 | 678 | 289 |
| 532 | 86 | 606 | 197 | 680 | 293 |
| 534 | 92 | 608 | 230 | 682 | 299 |
| 536 | 98 | 610 | 268 | 684 | 304 |
| 538 | 109 | 612 | 304 | 686 | 308 |
| 540 | 113 | 614 | 348 | 688 | 314 |
| 542 | 117 | 616 | 388 | 690 | 324 |
| 544 | 117 | 618 | 418 | 692 | 339 |
| 546 | 117 | 620 | 441 | 694 | 363 |
| 548 | 113 | 622 | 445 | 696 | 398 |
| 550 | 107 | 624 | 440 | 698 | 443 |
| 552 | 103 | 626 | 426 | 700 | 498 |

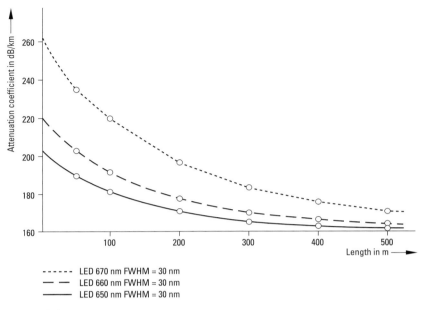

**Fig. 4.4**
Dependence of the attenuation coefficient on the length of the optical fiber using LEDs with different peak wavelength as transmitters

**Table 4.3**
Change in the attenuation coefficient dependent on the change in the peak wavelength of the transmitter for a plastic optical fiber of 50 m in length

| Change of peak wavelength of transmitter in nm | Increase in attenuation coefficient in dB/km |
|---|---|
| from 650 to 660 | 16 |
| from 660 to 670 | 41 |
| from 670 to 680 | 53 |

### 4.1.4 Launching conditions

Apart from the filter effect, launching conditions also have an influence on the result of the attenuation measurement. The launching conditions refer to the distribution of the input power launched in the fiber across the modes that are capable of propagation.

## Uniform mode distribution (UMD)

By means of homogenous launching of the light into the optical fiber across the entire core and the numerical aperture (full excitation), each excited mode initially carries the same energy (uniform mode distribution). As they pass along the optical fiber, however, the individual modes are attenuated to varying degrees. This is due to the fact that high order modes travel further in the fiber than low order modes (see Section 1.2.3). High order modes are thus subject to greater attenuation and their light power changes. In addition, an exchange of energy takes place between the individual modes, which affects the uniform mode distribution. This "mode coupling" is mainly a result of inhomogeneities in the core material, as well as fluctuations in diameter and unevenness of the core-cladding interface caused by the manufacturing process.

## Equilibrium mode distribution (EMD)

Beyond a certain distance, a steady state is reached, whereby the energy distribution over the modes remains constant. The measurement thus no longer depends on the distance. In addition, the equilibrium mode distribution (EMD) is now independent of the respective launching conditions. Thus in order to obtain reproducible test results, launching using EMD is necessary.

## Possibilities for implementing EMD

• Launching fiber

As already explained, EMD is reached beyond a certain distance in every optical fiber. It is possible therefore to connect a launching fiber of sufficient length to achieve the desired effect. In practice, however, this method is not advisable, as the fibers have to be unnecessarily long (up to several kilometers for glass fibers). With plastic fibers a length of about 30 or 40 m is sufficient. The disadvantage however is the high attenuation of the launching fiber and the possible filter effect.

• 70% launching

If the light launched into the fiber is limited to only 70% of the core diameter and the numerical aperture by means of lenses and apertures, high order modes are excluded from the start. In this way a mode distribution can be achieved that roughly corresponds to EMD.

• Mode scrambler

In most cases a mode scrambler is now used. This brings about the mode coupling effect mentioned above. To do this, mechanical perturbations are introduced by external effects (e.g. by bending around a small radius).

A mode scrambler may consist of two cylinders around which a plastic optical fiber is wound in a figure-of-eight configuration (see Fig. 4.5).

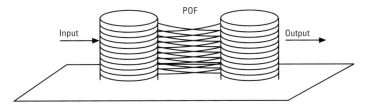

**Fig. 4.5**   Mode scrambler

Cylinders of 42 mm diameter and placed 3 mm apart are normally used, around which the fiber is wound 10 times. Consideration must be given to the fact that the peak wavelength at the end of the mode scrambler is shifted due to the filter effect. The attenuation of such a mode scrambler is about 8 dB.

In order to test how well this steady state is achieved, however, the near and far field distribution must be measured, although this will not be described in any further detail here.

Experience indicates that it is advisable to connect two mode scramblers in series. The principle of the optical attenuation characteristics, relative to the different excitation conditions, is represented in Fig. 4.6.

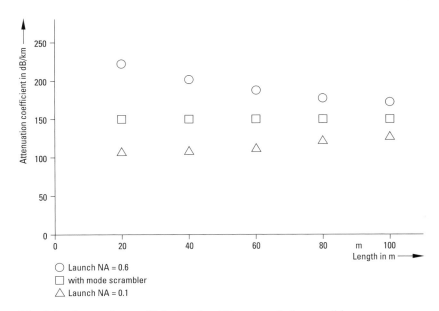

**Fig. 4.6**   Attenuation coefficient under different excitation conditions

## 4.1.5 Measurements in practice

The considerations so far have indicated that the attenuation measurement on plastic optical fiber is dependent first and foremost on its length and the excitation conditions. In order to achieve comparable and reproducible measurement results, these parameters must therefore be defined precisely. In practice, these requirements presents extreme difficulties, as the available LEDs exhibit marked scattering of the peak wavelength due to the way they are manufactured. In addition, the peak wavelength is temperature-dependent (see Section 8.2.1).

If an LED is nevertheless used as the source, the following parameters must be defined and observed:

▷ Peak wavelength of the LED,

▷ Full width at half maximum of the LED,

▷ Ambient temperature,

▷ Test set-up and procedure, in particular the definition of the launching conditions,

▷ Length of the optical fiber under test.

Another possibility is to use a monochromatic source as a transmitter (e.g. laser, that in this case can be considered as virtually monochromatic), which removes the dependence of the attenuation coefficient on distance. Using such a test arrangement, it is possible to test very great distances (>500 m) on the test unit. This is of particular interest for manufacturers of plastic optical fiber and cables and offers the user security with regard to the quality of the optical attenuation.

When using a monochromatic source, the following parameters must be defined:

▷ Peak wavelength and full width at half maximum of the LED

▷ Test set-up and procedure, in particular the definition of the launching conditions

The ambient temperature is only to be considered if the peak wavelength of the monochromatic source shows a temperature drift. A test set suitable for this purpose is illustrated in Fig. 4.7.

For the transmitter, a semiconductor laser with a peak wavelength of 651 nm and a full width at half maximum of 1 nm is used. The peak wavelength must be kept constant by means of appropriate cooling. An optical power of +5 dBm is launched into a plastic optical fiber (1 mm fiber). The receiver has a sensitivity of −108 dBm. Using this test unit, plastic optical fiber with a length of up to about 700 m, depending on fiber quality, can be reliably measured.

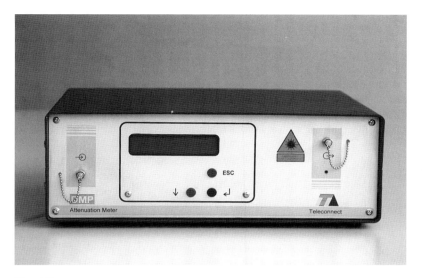

**Fig. 4.7**
Test set for measurining attenuation of plastic optical fibers
(photo: Teleconnect)

### Evaluation of the manufacturers' specifications

Manufacturers usually specify the optical attenuation as monochromatically measured, which is also advisable according to the above considerations. The user should note that the transmitters for data transmission are usually LEDs and that the attenuation of the plastic optical fiber is then about 30–40% higher, depending on the length of the fiber and the wavelength of the LED. Experience indicates that in each case it is advisable to obtain precise information from the manufacturer with regard to test frequency and test method. Some manufacturers also provide test certification on the delivery coil or in printout form.

## 4.2 Bandwidth measurement

Until now, there have been very different bandwidth specifications for 1 mm plastic optical fiber. According to IEC 60793-2, a bandwidth of ≥ 10 MHz over 100 m is demanded. This requirement is also met by most fibers currently available. The most recent efforts to use plastic optical fiber additionally for applications with data rates of up to 155 Mbit/s, as well as the associated development of low NA and graded index fibers on a polymer basis, are an incentive to consider this property and its measurement in greater detail.

**Table 4.4**
Bandwidth of step index plastic optical fiber with different NA of the fiber and the launching system

| NA of the fiber | NA = 0.47 | NA = 0.47 | NA = 0.31 |
|---|---|---|---|
| Launch-NA | 0.10 | 0.65 | 0.65 |
| Bandwidth-distance product in MHz · 100 m | ≈ 180* | ≈ 45* | ≈ 110** |

\* [4.1]
\*\* [4.2]

Just like the attenuation measurement, these measurements are dependent on the launching conditions. Measurements with a small numerical aperture (NA) of the source or receiver compared to the fiber produce higher bandwidths than measurements where the NA of the excitation system is the same or greater than the NA of the fiber. Table 4. 4. lists a selection of bandwidth measurements that have been carried out so far.

As already explained in Section 1.2.5, a pulse launched into an optical fiber becomes wider due to dispersion. At the same time, this pulse becomes more flat – the amplitude diminishes.

### 4.2.1 Measurement in the time domain (pulse response)

The pulse broadening can be measured by comparing it with a short reference fiber (about 2 m). To do this, a short light pulse is first launched into the optical fiber to be measured and then into the reference section. The pulse at the end of the optical fiber is amplified and forwarded to the input of the oscilloscope. By integrating the input pulse $g_1$ $(t)$ (measured on the reference section) and the output pulse $g_2$ $(t)$, the effective pulse widths $T_1$ and $T_2$. The effective pulse broadening $T_{eff}$ is calculated from

$$\Delta T_{eff} = \sqrt{T_2^2 - T_1^2} \qquad (4.11)$$

The bandwidth $B$ is approximated, assuming a Gaussian-shaped pulse:

$$B \approx \frac{0,375}{\Delta T_{eff}} \qquad (4.12)$$

To determine the bandwidth more precisely, a Fourier transformation of the pulses into the frequency domain must be performed. The bandwidth of the optical fiber is determined as that modulation frequency in which the amount of the frequency response G $(\omega)$ is 0. 5. How should this be interpreted? We saw at the start that the amplitude of the pulse declines as the frequency increases. The frequency response G $(\omega)$ is defined as:

$$G(\omega) = \frac{P_2(\omega)}{P_1(\omega)} \qquad (4.13)$$

This is normalized to the frequency response at the modulation frequency 0 Hz. The bandwidth then corresponds to the modulation frequency, at which the amplitude of the pulse has dropped to 50% compared with the value at frequency 0 Hz. In the logarithmic representation, the 50% drop represents a value of 3 dB (see Formula 1.4).

By measuring in the time domain with subsequent Fourier transformation, it is therefore also possible to determine the pulse response.

### 4.2.2 Measurement in the frequency domain

With this procedure the amplitude is measured relative to the frequency. The amplitude of the transmitter is modulated using a swept-frequency generator at a continuously rising frequency $\omega$. The power of the modulated signal is measured at the output of the optical fiber. The frequency response is determined in accordance with Formula 4.13, producing the bandwidth, as described above.

# 5 Buffered plastic optical fibers

## 5.1 Design

The buffered optical fiber is a widely used construction with which the fiber can be protected against external effects and damage, whereby a protective buffer with a typical wall thickness of 0.6 mm is extruded over the 1 mm fiber (Fig. 5.1).

Apart from the simplex (single) fiber, a duplex buffered fiber (see Fig. 5.1 b) can also be used.

The fiber code (compliant with DIN VDE 0888 Part 4) is explained below using the example of a fiber buffered with polyethelene (PE):

*Code*

I – V2 Y 1 P 980/1000 150 A 10

I – Indoor cable

V – Tight buffered

2 Y – Protective Polyethylene buffer

1 P – One plastic optical fiber with step index profile

980/1000 – Core/cladding diameter in μm

150 – Attenuation coefficient in dB/km

A – Wavelength 650 nm

10 – Bandwidth-distance product 10 MHz · 100 m

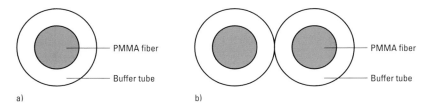

a)                                        b)

**Fig. 5.1**   a) Simplex buffered plastic fiber; b) Duplex buffered plastic fiber

## 5.2 Manufacture

The protective buffer is applied in a continuous process. An extruder is positioned after the take-off spool which applies the protective buffer material to the fiber (Fig. 5.2). The extrusion temperature depends on the material used in each case and is within the range of 160–250 °C. After buffering, the fiber passes through a cooling basin.

With the aid of a diameter checking device beyond the extruder, the diameter is monitored during manufacture. Particular attention must be paid during the process to the temperature, pressure and the tensile forces that arise. These parameters can influence the attenuation, geo-metrical form of the fiber, as well as the orientation of the polymers (see Section 3.3) and have a detrimental effect on the quality of the fiber. At the end of the system, the fibers are either wound onto drums or laid in horizontally stored trays or containers. As the fiber can only withstand tensile forces of up to 5 N, the take-off and winding drums must both be driven in order to compensate as far as possible for the tensile forces that occur in the manufacturing process. Typical raw lengths of such a fiber are between 2 and 5 km.

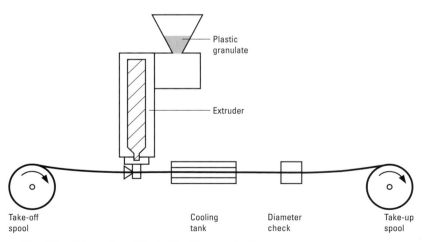

**Fig. 5.2** Schematic of the buffer fiber production

**Table 5.1** Properties of the commonest simplex plastic optical fiber buffer tubes

| Buffer tube material | Maximum operating temperature of buffer in °C | Typical attenuation in dB/km at 650 nm monochromatic | Typical attenuation in dB/km at 660 nm LED; length 50 m | Tensile strength in N (continuous load) | Minimum bending radius in mm | Crush resistance in N/cm | Halogen free | Flame resistance |
|---|---|---|---|---|---|---|---|---|
| Polyethylene (PE) | +80 | 150 | 210 | 5 | 30 | 10 | yes | poor |
| Polyvinylchloride (PVC) | +80 | 150 | 210 | 5 | 30 | 10 | no | good |
| Cross-linked polyethylene (VPE) | +115 | 300 | 450 | 10 | 30 | 10 | yes | poor |
| Polyamide (PA) | +85 | 150 | 210 | 10 | 30 | 15 | yes | poor |
| Chlorinated polyethylene (CPE) | +80 | 150 | 210 | 5 | 30 | 10 | no | good |
| Ethylenevinylacetate copolymer (EVA) | +80 | 150 | 210 | 5 | 30 | 10 | yes | good |

# 5.3 Properties

## Technical properties

Depending on the requirements profile, a host of different plastics can be used for the buffered fibre which ultimately have a decisive influence on the properties of the fiber. The standard fibers and their properties are listed in Table 5. 1. The corresponding test procedures and parameters for the individual properties are explained in Chapter 7.

The maximum values for tensile strength specified in Table 5.1 apply to continuous loading. For momentary loading, the specified limit can be exceeded. Typical limit for momentary loading is in the region of 60 N at +23 °C, which does not damage the fibers or change their properties.

The specifications for the minimum bending radius vary widely. For the user it is on the one hand crucial to know what increase in attenuation occurs at each radius and on the other hand, at what point this attenuation increase becomes irreversible, i.e. destruction of the fiber occurs. At the specified radius of 30 mm, an attenuation increase of about 0.1 dB is to be expected. A smaller radius can be tolerated momentarily without the fiber being destroyed. This only causes a momentary rise in the attenuation which disappears as soon as the radius is increased again. Momentarily permissible radii are usually included in the manufacturer's specifications.

If the buffered fibres are subjected briefly to a transverse compressive stress, no measurable changes in attenuation occur for stresses up to 10 N/cm. Under greater loads, reversible changes to the attenuation are noticeable. For example, an increase of 0.6 dB in the attenuation could be measured on a PE fiber subjected to a load of 195 N/cm for three minutes. After removing the load, the attenuation returned to the initial status [5.1].

The halogen-free property specified in Table 5.1 is checked in accordance with DIN VDE 0472 Part 813 as corrosiveness of the combustion gases. If flame resistance as specified in DIN VDE 0472 Part 804 is demanded in addition to freedom from halogens, then special materials such as those based on ethylene vinylacetate must be used.

It should also be mentioned at this point that a PVC fiber can readily be secured in a connector with adhesive, whereas PE can only be connected to other materials with great effort, using dual-component adhesives.

## Dimensions

The external diameter of the 1 mm plastic optical fiber with buffer is normally 2.20 ± 0.07 mm. The duplex fiber measures 2.2 × 4.4 mm (height × width).

**Coding**

Besides the manufacturer's labeling, the fiber code should also be printed on the buffer.

So far there is no standardized or firm ruling about color coding of the buffer. Many users would prefer the buffer to have bright or luminous color, e.g. red, which has at least two advantages: First, the plastic optical fiber is recognized more quickly in use and second, it causes the technicians to take greater care when handling plastic optical fibers. Most buffers, however, are black and PVC is usually supplied in gray.

When choosing a color other than black, it is essential to remember that extraneous light can penetrate the buffered fibre. This external light can affect the function of the receivers which operate with various degrees of sensitivity. If this is taken into account when manufacturing the buffer and if a double coating is applied (the base coat being black), such interference can be avoided.

## 5.4 Applications

In environments subject to minimal external stress (e.g. switchgear cabinets) and over short distances, a fiber with an external diameter of 2.2 mm is most commonly used. In addition, the fiber is used in the cables described in Chapter 6.

# 6 Plastic optical fiber cable design

As the fundamental principles of plastic optical fibers have been explained in Chapter 5, we will now consider more complex cable constructions that are suitable for special applications.

## 6.1 Non-stranded cables

If a plastic optical fiber cable consists of just one or two core fibers, they are usually not stranded. Only cables with higher numbers of fibers (multifiber cables) are stranded.

### 6.1.1 Simplex cable

The typical POF simplex cable consists of a plastic optical fiber buffer that is surrounded by strength elements and an additional outer sheath. The strength members are made of special yarns which are either woven around the buffered fibre or simply laid in parallel with it (see Fig. 6.1)

The code (consistent with DIN VDE 0888 Part 4) is explained below using the example of a polyurethane-sheathed cable:

*Code*

I – V11 Y 2 Y 1 P 980/1000 160 A 10

I – Indoor cable

V – Tightly buffered

11 Y – Polyurethane (PUR) sheath

2 Y – Protective polyethylene buffer

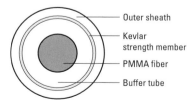

**Fig. 6.1**   Plastic optical fiber simplex cable

1 P – One plastic optical fiber with step index profile

980/1000 – Core/cladding diameter in mm

160 – Attenuation coefficient in dB/km

A – Wavelength 650 nm

10 – Bandwidth-distance product 10 MHz · 100 m

### 6.1.2 Duplex cable

The standard type of duplex cable consists of two parallel fibers with tensile strength members running in the gap in between them. In most cases they are then wrapped in a plastic film which facilitates removal of the outer sheath (see Fig. 6.2). For easier identification, the two fibers are of different colors or are identified by printing on the buffer. The disadvantage of using a duplex cable instead of two simplex cables is that high tension and pressure forces can be exerted on the fibers when the cable is bent. An increase in the attenuation cannot be ruled out in this case.

In another duplex construction, the twin cable, the two fibers are individually buffered and sheathed with their own individual strength members (Fig. 6.3). The advantage of this construction is in the possibility of extending the strength members as far as the connector where they can be attached.

The code (compliant with DIN VDE 0888 Part 4) is explained below using the example of a PVC-sheathed twin cable:

*Code*

I – VYY 2P 980/1000 160 A 10

I – Indoor cable

V – Tight buffered

Y – PVC sheath

Y – PVC buffer

2 P – Two plastic optical fibers with step index profile

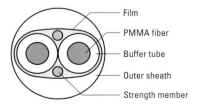

**Fig. 6.2**  Duplex cable

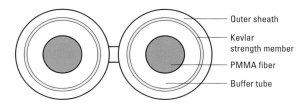

**Fig. 6.3** Twin cable

980/1000 – Core/cladding diameter in mm
160 – Attenuation coefficient in dB/km
A – Wavelength 650 nm
10 – Bandwidth-distance product 10 MHz · 100 m

### 6.1.3 Manufacture

The outer sheath is applied in a continuous process similar to that in the buffered fibre manufacture (see Section 5.2). A yarn strandex for introducing the strength member yarns is positioned ahead of the outer sheath extrusion unit.

In this manufacturing process too, the temperature, pressure and tensile forces can have a negative effect on the optical attenuation of the plastic optical fiber. As a rule, an attenuation coefficient rise of 10 dB/km is to be expected as a result of this manufacturing stage. Precise details will be supplied by the manufacturer.

### 6.1.4 Properties

Typical properties of a simplex cable are summarized in Table 6.1 together with its dimensions.

The outer sheath is usually made of polyurethane (PUR) or PVC. PUR is particularly resistant to abrasion and oil and is therefore frequently used for industrial purposes.

The tensile strength limits specified in Table 6.1 apply to continuous loading.

### 6.1.5 Applications

Cables with strength members are mandatory wherever cables have to be laid in the field either in cable ducts or on trays, from equipment to equipment or between switchgear cabinets. Thanks to the greater tensile strength and the better protection of the optical fiber afforded by the additional outer sheath, damage to the fiber during installation can largely be ruled out.

55

**Table 6.1**  Typical properties and dimensions of a simplex cable

| Type | Simplex cable with strength member |
|---|---|
| External diameter in mm | 3.6 |
| Operating temperature in °C | −40 to +80 |
| Typical attenuation  in dB/km at 650 nm monochromatic | 160 |
| Typical attenuation  in dB/km at 660 nm LED; length 50 m | 230 |
| Tensile strength in N (continuous load) | 100 |
| Minimum bending radius in mm | 50 |
| Crush resistance in N/cm | 20 |

## 6.2  Stranded cables

### 6.2.1  Design

Cables with several optical fibers, and also cables with a combination of fibers and copper wires, are stranded. Additional strength members can also be integrated. The central element can also be used as a strength member. In addition, blank elements, i.e. buffers without any fibers or copper wires, can be accommodated in the cable core. The stranded cables are usually wrapped in a tape or plastic film for further protection. The entire cable configuration with any wrapping is called a cable core and is enclosed in the outer sheath in another manufacturing stage. Fig. 6.4 shows such a construction using the example of a hybrid cable which combines plastic optical fibers with copper conductors.

The code (compliant with DIN VDE 0888 Part 4) describes the copper conductors in addition to the plastic optical fibers (see Appendix). The following code applies to a hybrid cable with three plastic optical fibers and three copper conductors:

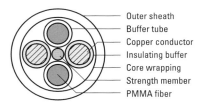

**Fig. 6.4**
Hybrid cable with two plastic optical fibers and two copper conductors

*Code*

I – V11Y2Y 3P 980/1000 250 A10 +3x1 FF-Cu300/500 V

I – Indoor cable

V – Tight buffered

11 Y – Polyurethane sheath

2 Y – Polyethylene buffer

3 P – Three plastic optical fibers with step index profile

980/1000 – Core/cladding diameter in mm

250 – Attenuation coefficient in dB/km

A – Wavelength 650 nm

10 – Bandwidth-distance product 10 MHz · 100 m

3x1–3 copper conductors each with a cross section of 1 mm$^2$

FF-Cu – Extra-finely stranded copper conductor

300/500 V – Rated voltage $U/U_0$

## 6.2.2 Manufacture

During the stranding process, care should be taken that the tension, torsion and pressure forces on the buffered fibres are kept to a minimum, as these have a decisive influence on the optical attenuation of the optical fiber. This requires the use of special stranding machines. There are two common types of stranding: helical stranding and reverse lay (SZ) stranding. In the first method, all elements are stranded in the same direction with a constant angle to the longitudinal axis of the cable. The take-off spools for the stranding elements are positioned around the circumference of a rotor and twisted around the stranding axis. The finished binder tape is drawn off perpendicular to the rotor. The reverse turning is performed by turning the take-off spools themselves, in the opposite direction to the rotor, and serves to reduce the torsional tension in the stranding elements that is generated by the stranding process.

The SZ stranding method, on the other hand, permits a continuous and thus efficient manufacturing process, as the take-off spools for the stranding elements are not located on a rotor. The rotor necessitates that the stranding process is halted when the take-off spools have to be refilled. In the case of SZ stranding, the direction of the stranding is reversed after a certain number of turns. This means that the stranding elements first describe an S-shape and then, following the change of direction, a Z-shape (see Fig. 6.5).

Within the stranding, an individual stranding element describes a helical line, the pitch of this line after a full turn of 360° being termed the *lay*

a) S-direction                      b) Z-direction

**Fig. 6.5**   SZ-stranding a) S-rotation; b) Z-rotation

*length S*. "High flexibility" cables designed for installations permanently in motion have a shorter lay length than cables intended for fixed installations.

Because of stranding, the stranding elements must be longer than would be necessary if they were parallel to the longitudinal axis as in non-stranded cables. In addition, the stranding is continuously curved around a specific bending radius. Both these effects have an influence on the optical attenuation of the optical fiber in the finished cable. But how great is this influence? The length $L$ of a stranding element can be calculated according to Formula 6.1, where the stranding radius $R$ refers to the distance between the axis of the cable and the center of a stranding element. $S$ represents the lay length.

$$L = S \cdot \sqrt{1 + \left( \frac{2\pi R}{S} \right)^2} \tag{6.1}$$

To determine how long the individual stranding element s are in a cable of specified length, the stranding excess $Z$ is used, which is calculated according to the following formula:

$$Z = \frac{L - S}{S} \cdot 100\% = \left( \sqrt{1 + \left( \frac{2\pi R}{S} \right)^2} - 1 \right) \cdot 100\% \tag{6.2}$$

The attenuation coefficient of the entire cable increases with the same percentage compared with a non-stranded optical fiber, as the stranding excess.

Example:

For a hybrid cable with two optical fibers and two copper conductors 500 m long, a lay length of $S = 60$ mm and a stranding radius $R = 2.16$ mm, the stranding excess $Z$ is calculated as follows:

$$Z = \left( \sqrt{1 + \left( \frac{2\pi \, 2{,}2}{60} \right)^2} - 1 \right) \cdot 100\% \approx 2{,}6\%$$

This means that a single stranding element is 2.6% longer the overall cable. In our example of the 500 m long cable it is about 12.5 m longer.

For an assumed attenuation coefficient of 160 dB/km for the plastic optical fiber, this stranding excess of 2.6% raises the attenuation coefficient to ≈ 164 dB/km.

A stranded plastic optical fiber cable therefore will have a slightly higher attenuation coefficient than a non-stranded cable.

The radius of curvature $r$ of the screw line, or helix, described by the stranding element can be calculated as follows:

$$r = R \left( 1 + \left( \frac{S}{2 \pi R} \right)^2 \right) \tag{6.3}$$

The data from the example above produces a bending radius $r = 44.4$ mm. From a mechanical viewpoint, this bending radius is permissible for a plastic optical fiber. A bending radius of this magnitude does not have a significant effect on the optical attenuation (see Section 7.3).

Mechanical loads, such as transverse compressive stress and torsion, can have a considerably greater influence on the attenuation of the stranded plastic optical fiber. The manufacturer will provide information on this in each specific case.

After stranding, the core is wrapped and fitted with the outer sheath. The materials used for this are again PVC, PUR, EVA and others, depending on the application.

### 6.2.3 Properties

Due to the different compositions, each cable has different properties. As an example, important properties of two different stranded cables are listed in Table 6.2.

**Table 6.2** Typical properties and dimensions of stranded cables

| Code | I-VYY<br>6P980/1000 250A10 | I-V11Y2Y<br>2P980/1000 250A10+2x1FFCu |
|---|---|---|
| External diameter D of cable in mm | 9.4 | 7.7 |
| Operating temperature °C | −20 to +80 | +5 to +70 |
| Typical attenuation in dB/km at 650 nm monochromatic | 250 | 250 |
| Typical attenuation in dB/km at 660 nm LED; length 50 m | 280 | 280 |
| Tensile strength in N | 7 | 15 |
| Minimum bending radius in mm | 8 x D | 8 x D |
| Application | for permanent installation | for high-flexibility applications |

Compared to glass optical fibers, plastic optical fibers are less sensitive to bending, particularly repeated bending. In spite of the simple structure, therefore, highly flexible cables can be manufactured that are also suitable for use in moved applications. These cables are specifically specified by manufacturers for these applications (see Sections 7.4 through 7.6).

### 6.2.4 Applications

Multifiber cables are used wherever data is to be transmitted in parallel on several optical fibers. In many cases, the cost of installation can be significantly reduced if multifiber cables are used instead of numerous individual cables. Hybrid cables are used wherever a voltage is required, e.g. as a supply to telemetry devices, in addition to the data transmission.

#### Coding on the outer sheath

Manufacturers are free to choose the color of the outer sheath, frequently opting for red. A growing number of users, however, particularly in the field of mechanical engineering, are demanding that cable colors be standardized. There is a request that bus cables – for which plastic optical fiber cables are often used – be colored purple as standard to make them easier to identify after installation.

In addition to the manufacturer-specific designations, the labeling on the outer sheath carries the coding and in many cases have also a meter marking.

## 6.3 Notes on field installation

When laying plastic optical fibers in the field, only cables with strength members i.e. no buffered fibres should be used. What is more, the outer sheath of such cables offers additional protection against many external mechanical loads. When laying the cable, special care should be taken not to exceed the minimum bending radius, as otherwise the attenuation in the plastic optical fiber increases. Destruction of the fiber due to an excessively tight bending radius is however unlikely, unless the cable is kinked. From time to time, the manufacturers specify a greater bending radius for cable-laying than for operation. This can be explained by the fact that, during laying, additional torsional and tensile forces can act on the fiber in the area of the bend. As a precaution, therefore, a higher bending radius should be selected during cable laying.

The cables can be held in place with cable clips or fixings, but the cable should be free to move in the area of the bend, with cable fixings being applied before and after the bend. The cable should not be squashed by the cable clips or fixings.

The distances between the devices to be connected are often not know prior to laying the cable. The meter marking on the cable helps to determine the actual length of the cable once laid. On one hand this is necessary for the functionality of the overall system and on the other hand it assists in troubleshooting for potential faults.

When laying plastic optical fibers in trailing chains (see Section 7.5), special care must be taken to insure the cable is laid without twisting or tension. The cable must be free to move in the chain. Furthermore, any specific instructions by the manufacturer regarding laying in trailing chains must be observed.

# 7 Test procedures

Unless otherwise specified, the tests described below are performed under standard environment conditions for tests in agreement with 5.3 of IEC 60068-1.

## 7.1 Tensile strength

The tensile strength of plastic optical fibers is tested in accordance with DIN EN 187000, Test Procedure 501 A and B or IEC 60794-1-E1. The objective of the test is to examine, on the one hand, the behavior of the optical attenuation and, on the other hand, the fiber elongation of the cable as a function of the tensile force. From this, the maximum permissible tensile force for the cable is derived.

**Test sample**

As a rule, a 100 m long sample of the plastic optical fiber cable is tested.

**Test equipment and performance**

A suitable test arrangement is shown in Fig. 7.1. The clamping devices must be designed in such a way that they do not affect the test results.

Once the cable sample is in place, the tensile force gradually increased, normally at a rate of 100 mm/min. Every change in the attenuation and elongation is recorded.

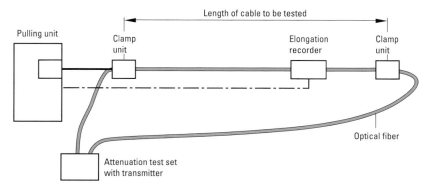

**Fig. 7.1**   Tensile test arrangement

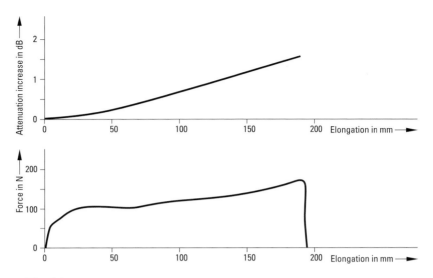

**Fig. 7.2**
Schematic of the behavior of a optical buffered fibre with PE sheath at T = 23 °C; $\lambda$ = 660 nm, pulling rate = 100 mm/min; sample length = 100 m

## Results

The attenuation and/or the elongation of the test sample should not exceed the value given in the design specification. The change in attenuation and/or the elongation is represented as a function of the tensile force, whereby the wavelength, temperature and speed are specified as the tensile force is increased. The behavior of a plastic optical fiber is shown in Fig. 7.2.

## 7.2 Static bending

### 7.2.1 Bending around 90°

The following test is used to determine the rise in optical attenuation caused by static bending around an angle of 90°. This value is of particular importance for planning systems and when installing cables.

### Test equipment and performance

As shown in Fig. 7.3, the cable is bent around a test mandrel. The attenuation is measured as the cable is being bent. A schematic of the behavior of a plastic optical fiber as a function of the bending radius, is shown in Fig. 7.4.

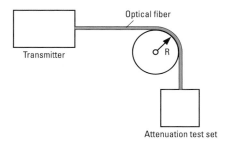

**Fig. 7.3**    Arrangement for the bending test around an angle of 90°

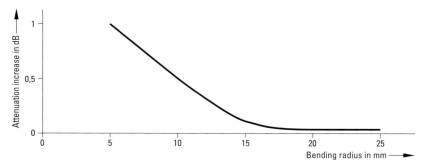

**Fig. 7.4**
Rise in attenuation of a plastic optical fiber bent around 90° as a function of the
bending radius

## 7.2.2 Bending around 360°

This test is performed in accordance with DIN EN 187000, Test Procedure 513 or IEC 60794-1-E1. This test measures the rise in the optical attenuation of a plastic optical fiber cable that is bent repeatedly around
360°.

### Test equipment and performance

A suitable test arrangement is shown in Fig. 7.5. Once connected to the
transmitter and attenuation test unit, the connectorized cable is wound
tightly around the mandrel. The increase in attenuation is recorded after
each winding or after the complete number of windings. The cable is
then unwound, laid straight and the attenuation is measured again. The
complete winding and unwinding process represents one cycle. The number of cycles to be performed is defined in the design specification of the
cable in question.

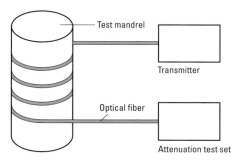

**Fig. 7.5**   Test equipment for static bending around 360°

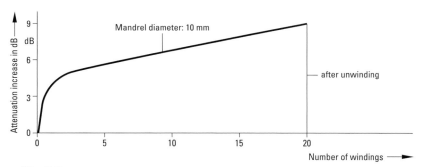

**Fig. 7.6**
Behavior of a plastic optical buffered fibres on repeated bending around a test
mandrel

### Results

The increase in attenuation  after winding and after completion of one
cycle must not exceed the specified values. No signs of cracking in the
cladding must be visible to the naked eye. The fiber must not break during the test.

As can be seen in Fig. 7.6, the attenuation increase has a non-linear relationship to the number of windings. The majority of modes capable of
propagation are coupled out in the first few windings. As the number of
windings increases, a diminishing number of propagating modes are coupled out, thus resulting in the non-linear behavior.

## 7.3 Repeated bending

This test is performed in accordance with DIN EN 187000, Test Procedure 507 or IEC 60794-1-E6 and is used to examine the resistance of the plastic optical fiber to repeated bending. One outstanding feature of plastic optical fibers is their excellent bending behavior. For this reason, this test, like the examinations of alternate bending resistance and trailing chain suitability, is of particular interest.

**Test equipment and performance**

A suitable test arrangement is shown in Fig. 7.7. With this test equipment the sample is bent through a total of 180°, once through 90° to the right and then once through 90° to the left. A cycle begins in the central vertical position. Within two seconds the plastic optical fiber must be bent first to the right, then to the left and finally returned to the starting position. During bending, the sample is subjected to a tensile force. As a rule, a 200 g weight is hung on the fiber. The bending radius for buffered fibres is usually 40 mm, while for cables it is usually ten times the external diameter.

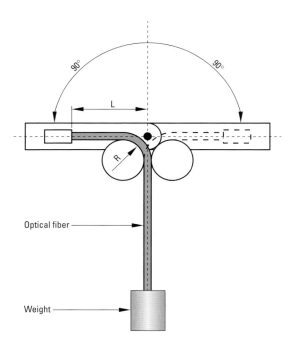

**Fig. 7.7**   Test equipment for repeated bending

Once the sample has been fixed in place, the required number of bending cycles are carried out. Plastic optical fibers are usually tested with 50000 cycles.

**Results**

The fiber must not break during the test. The increase in attenuation should remain below the limit specified in the design specification. In the case of a plastic optical fiber with a PE plastic sheath, the value for the above requirements (at 660 nm) is typically below 0.5 dB.

## 7.4 Alternate bending test

This test is performed in accordance with DIN EN 187000, Test Procedure 509 or IEC 60794-1-E8 and is used to examine the resistance of the plastic optical fiber to alternate bending.

**Test equipment and performance**

A weight is hung on each end of the cable. The cable is threaded over guide pulleys at each end to two further pulleys (marked "A" and "B" in Fig. 7.8) which are mounted on a moving trolley. There is a groove on each pulley to guide the cable and the retaining clamps are attached in

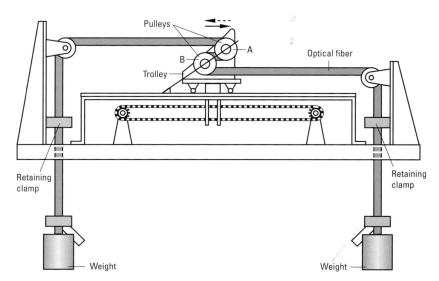

**Fig. 7.8**  Equipment for alternate bending test

such a way that the tensile load on the cable is always caused by the weight, from which the trolley is moving away.

The test conditions, such as the diameter of pulleys A and B, the weight and number of cycles must comply with the design specifications. Typically for plastic optical fibers, 200 g weights and pulley diameters of 40 mm are selected. After attaching the buffered fibre, the required number of alternate bending cycles is carried out (generally 10 000 for plastic optical fibers).

**Results**

In the case of a plastic optical buffered fibre with a PE plastic sheath, the increase in attenuation for the above requirements is typically below 0.5 dB.

## 7.5 Trailing chain test

*Trailing chain* cables or energy lead cables are being used increasingly in industry (see Fig. 7.9). Cables of this type facilitate the particularly reliable routing of energy and data cables and they are used wherever frequently moving machines or components have to be supplied with power and data.

In order to specify the plastic optical fiber for precisely this application, the trailing cable test is performed. The test takes place under realistic conditions in a trailing chain itself. The test parameters of chain radius, chain travel speed, travel distance and acceleration can vary widely and depending on the parameter values selected, widely varying results are obtained. For example, a trailing cable with a short travel path and a high acceleration can produce very different results than a cable with low acceleration and a long travel.

A host of investigations with different parameters have demonstrated, however, that plastic optical buffered fibres and cables are very well suited to applications of this type. For example, examinations with a chain radius of 77 mm, a travel path of 10 m, a speed of 2 m/s and an acceleration of 1 m/s$^2$ have shown that there is no increase in attenuation even after the fiber has been bent more than one million times [7.1]. Both PE buffered fibres and hybrid cables with two plastic optical fibers and two copper conducts have been tested.

## 7.6 Torsion

This test is performed in accordance with DIN EN 187000, Test Procedure 508 or IEC 60794-1-E7 and is used to examine the resistance of a plastic optical fiber cable to torsion.

**Fig. 7.9** Trailing chain cables in use

## Test equipment and performance

The cable is fitted with a connector at each end and placed in the test unit (see Fig. 7.10) which has one fixed and one rotating cable clamp. Prior to the test, the sample is connected to the optical transmitter and the attenuation test set. A torsion cycle consists of a certain number of turns in one direction, the same number of turns back to the starting position and then a specified number of turns in the opposite direction. The number of cycles, the size of weight and the length of the sample are to be obtained from the design specification. Plastic optical buffered fibres are typically tested with a sample length of 1 m with ± 10 times 360° turns per cycle, a weight of 200 g and a cycle time of 40 seconds.

## Results

The evaluation criteria are also stated in the design specifications. In the case of a PE buffer under the above conditions (1000 cycles and 23 °C) an increase in attenuation of about 0.2 dB is to be expected (at 660 nm).

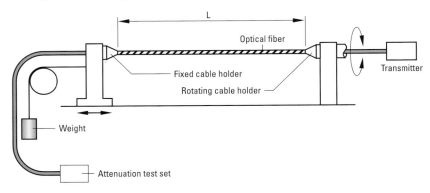

**Fig. 7.10**  Test equipment for a torsion test

## 7.7 Impact resistance

This test is performed in accordance with DIN EN 187000, Test Procedure 505 or IEC 60794-1-E4 and is used to examine the impact resistance of a plastic optical fiber cable.

**Test equipment and performance**

As shown in Fig. 7.11, a weight is allowed to drop vertically onto an impact plate which transfers the impact directly onto the test sample which is lying on a flat steel base.

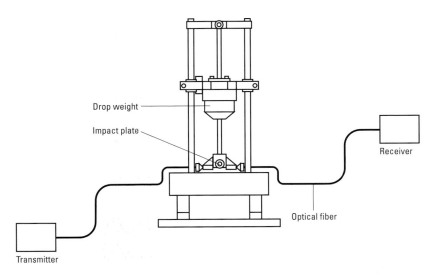

**Fig. 7.11**  Test equipment for impact test

The test conditions such as drop height, weight and number of impacts are to be found in the appropriate design specifications. Typically plastic optical fibers are tested with a drop weight of 1 kg and a drop height of 20 mm.

**Results**

In the case of a PE buffered fibre subjected to 200 impacts under the above conditions, an increase in attenuation of about 0.1 dB is to be expected (at 660 nm).

## 7.8 Transverse compressive stress

This test is performed in accordance with DIN EN 187000, Test Procedure 504 or IEC 60794-1-E3 and is used to examine the resistance of a plastic optical fiber cable to transverse compressive stress.

**Test equipment and performance**

The test equipment (see Fig. 7.12) comprises a firm steel base plate and a moveable steel plate on which the transverse compressive stress is exerted. The cable is positioned between the two plates so that no lateral movement is possible. The moving plate is designed in such a way that the transverse compressive stress is exerted evenly over a sample length of 100 mm. The pressure must be increased gradually, during which the change in attenua-

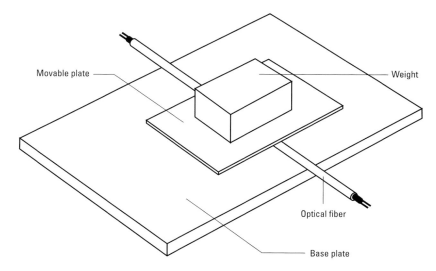

Movable plate

Weight

Optical fiber

Base plate

**Fig. 7.12**  Test equipment for transverse compressive stress test

tion is recorded. Once the compressive stress has been applied for a specific period, it is released again for a specific period. One cycle consists of one stress phase followed by one relief phase. The precise test conditions should again be obtained from the corresponding design specifications. As a rule, plastic optical fibers are subjected to a compressive stress of 2000N for three minutes. The relief period is typically six minutes.

**Results**

The changes in attenuation during the period of compressive stress and following the relief of this stress, must not exceed the limits given in the design specification. The typical behavior of a PE buffered fibre is shown in Fig. 7.13.

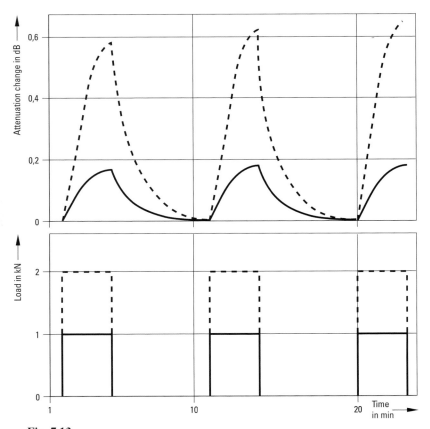

**Fig. 7.13**
Schematic of the behavior of a plastic optical buffered fibre with PE sheath, subjected to transverse compressive stress

## 7.9 Flame resistance and halogen-free properties

To afford better protection to human life and buildings in the event of a fire, there is a growing demand for cables that are flame resistant and halogen-free. The test procedure for examining the behavior of the cable when subjected to flames is described here. In addition, the smoke measurement can be made both quantitatively and qualitatively in the course of these tests. During a fire, halogen-bearing materials can give off halogens that combine with water to produce corrosive acids. For this reason, efforts are made to keep the cables halogen-free. If not only the flame resistance and smoke-tightness, but also the halogen-free property are specified for a cable, the cable can be labeled FRNC (flame retardant non corrosive). In some cases the designation LSOH (low smoke zero halogen) can be found, which in principle means the same as FRNC.

There are numerous procedures at both national and international level for testing the flame retardant properties of cables. Table 7.1 provides a summary of the major tests.

The test procedures for zero halogen are listed in Table 7.2 and those for smoke tightness in Table 7.3.

**Table 7.1** Overview of the important tests for flame resistance

| Standard | Arrangement of probes | Combustibility requirement | Smoke tightness requirement | complies with | to be superseded by |
|---|---|---|---|---|---|
| DIN VDE 0472 Part 804 Test Type A | vertical, single cable burn test | self extinguishing | none | IEC 60332-2 | – |
| DIN VDE 0472 Part 804 Test Type B | vertical, single cable burn test | self extinguishing | none | – | IEC 60332-1 |
| DIN VDE 0472 Part 804 Test Type C | vertical, multiple cable burn test | self extinguishing | none | – | IEC 60332-3 Cat. C |
| UL-1581 Par. 1160 | vertical, multiple cable burn test | self extinguishing | none | IEEE 383 | – |
| UL-1581 VW1 Par. 1080 | vertical, single cable burn test | self extinguishing, no burning drips | none | – | – |
| UL910 Steiner Tunnel | horizontal, multiple cable burn test | self extinguishing | yes | – | – |
| UL 1666 Riser | vertical, multiple cable burn test | self extinguishing | none | – | – |

**Table 7.2**  Tests for zero halogen

| Standard | Designation | Complies with |
|---|---|---|
| DIN VDE 0472 Part 815 | Halogen free | – |
| DIN VDE 0472 Part 813 | Corrosiveness of combustion gases | IEC 60754-2 |
| IEC 60754-1 | Halogen free | – |

**Table 7.3**  Tests for smoke tightness

| Standard | Designation | Complies with |
|---|---|---|
| DIN VDE 0472 Part 816 | Smoke-tightness | IEC 1034 |

As an example, the flame retardance test conforming to UL 1581 VW-1 is described below:

**Testing the flame retardance to UL 1581 VW-1, Paragraph 1080**

*Test equipment*

The sample (See Fig. 7.14) is held in the test equipment and a paper flag is attached at its top end. A distance of 254 mm must be maintained between the paper flag and the flame.

*Execution*

The flames must not burn as far as the paper flag during the test or continue burning more than one minute after five applications of the flame. Each application lasts 15 seconds followed by a 15-second pause. The cotton must not be ignited by dripping material.

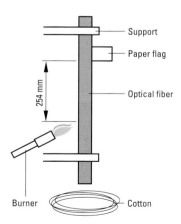

**Fig. 7.14**
Test equipment conforming to UL 1581 VW-1

74

# 8 Transmitting and receiving components

## 8.1 Physical principles

Suitable electrical-to-optical transducers (transmitters) are required at the beginning of the transmission link to convert electrical signals into optical ones. At the end of the link these signals are then converted back to electrical signals by means of corresponding optical-to-electrical transducers (receivers). The transmitters and receivers exploit the physical properties of semiconductor materials which allow the implementation of components with small dimensions and a long lifetime.

Atoms, molecules and solid-state components (e.g. semiconductor crystals) can absorb or emit energy in the form of electromagnetic radiation and interact with one another in this way. The fundamental processes that determine this interaction are emission, absorption and light propagation. They are closely linked with the atomic, molecular or solid-state structure. A decisive factor in understanding the interaction between light and material is the fact that electrons bonded in the atom only have a certain discrete energy value. The emission or absorption of light is linked with the transition of an electron from one energy stage to the next higher stage (absorption) or the reduction of the energy on the emission of a photon. This means that an atom can only absorb or emit light of quite specific wavelengths.

If the atoms in the molecules or solid-state component are closely linked with one another, this results in the "fanning out" of the originally sharply defined energy level into a host of closely adjacent levels, the energy bands. Due to the large number of atoms in the solid-state component, the possible energy values within a band are in effect continuous. The bands are separated from one another by energy gaps analogous to the relationships in the atom.

The electrons that are responsible for bonding the atoms and the molecules, are found in the valence band. Due to their bonding characteristics they are essentially localized, i.e. bonded to the atomic residue. The next higher band is the conduction band. The electrons in this band are in effect no longer bonded, but can move more or less freely in the solid-state element. They are the cause of the electrical conductivity.

Due to the transition from well-defined energy levels to the essentially broader bands, the spectral range of absorbing photon energies is correspondingly broad in solid-state elements. A prerequisite for a reduced radiation transition however is that be photon this must have a minimum energy, that is great equal to or greater than the energy gap between the bands. The energy gap for silicon, for example, is 1.2 eV. (electron volts).

If an electron "lifted" by absorption of the photon from the valence band into the conduction band, then this electron is absent from a bonding between the adjacent constituents of the semiconductor crystals. The "hole" created in this way can then be occupied by an neighboring electron, and the newly created hole by another neighboring electron. In this way, the hole can move through the semiconductor crystals and behaves like a freely moving positive charge.

The absorption of a photon in a semiconductor, therefore, always generates a pair of charge carriers that consists of one electron in the conduction band and one hole in the valence band. After a certain time, the lifted electron gives up its energy again, electron and hole recombine. This recombination may take place as a radiative transition, emitting a photon, or as a non-radiative transition, for example emitting the energy to the crystal lattice in the form of heat. The processes described are illustrated in schematic form in Figs. 8.1 and 8.2.

The band distance $E_g$ is calculated from the difference between the energy level of the valence band $E_V$ and the conduction band $E_L$:

$$E_g = E_L - E_V \tag{8.1}$$

In the equivalent photon energy $W_{ph}$ is then.

$$E_g = W_{ph} = hf = h\frac{v}{\lambda} \tag{8.2}$$

where h is the Planckian efficiency quantum, $f$ the frequency, $v$ the phase speed in the material and $\lambda$ the wavelength.

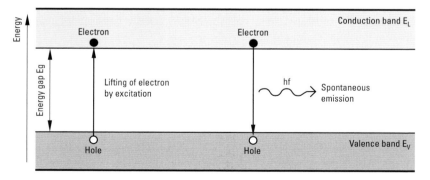

**Fig. 8.1**  Light emission with semiconductors (transmitter)

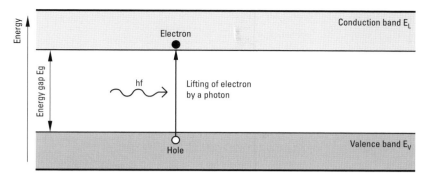

**Fig. 8.2**   Internal photoelectric effect (receiver)

Apart from the release of energy caused by spontaneous emission of a photon, an electron in the conduction band can be "stimulated" with a specific probability to recombine while emitting a photon. The photon emitted by stimulation then matches the photon that the emission has triggered in terms of energy (wavelength or frequency), phase and propagation direction of the radiation field. This stimulated emission is the basis for the function of the laser (see Fig. 8.3).

## 8.2 Transmitters

### 8.2.1 Fundamental properties

The following demands are made upon the transmitting components:

▷ High reliability

▷ High efficiency

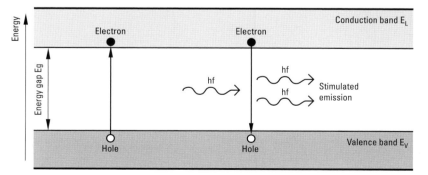

**Fig. 8.3**   Light emission in the laser

▷ Operation at normal temperatures

▷ High output performance

▷ Small active zone, good coupling to the optical fiber

▷ Simple, broadband modulation capability

▷ Low costs

▷ Narrow spectral emission

▷ Good linearity (important for analog technology)

Two types of transmitter element are used in optical transmission technology: light emitting diodes (LED) and laser diodes (LD). The different properties and applications of LED and LD are listed in Table 8.1.

In the red wavelength band the use of LEDs based on gallium arsenide phosphide (GaAsP) and gallium aluminum arsenide (GaAlAs) is preferred today as a semiconductor material in the case of transmitter diodes used for plastic optical fiber transmission. Within the scope of newer developments, aluminum indium gallium phosphide (AlInGaP) is used with a peak wavelength of 650 nm. Gallium indium phosphide is used for laser diodes, while in the green wavelength band, diodes based on gallium phosphide (GaP) are used. By comparison, gallium arsenide (GaAs) is used for glass optical fibers, the emission wavelength being determined by the band gap of about 0.9 μm.

The values between 0.8 μm and 0.9 μm (1$^{st}$ optical window) are achieved by the use of gallium aluminum arsenide. Even greater emitted wavelengths of 1.3 m (2$^{nd}$ optical window) or 1.55 μm (3$^{rd}$ optical window) can be achieved by the use of gallium indium arsenide phosphide (GaInAsP).

## Light emitting diodes (LED)

The basis for the function of the light emitting diodes or the laser diodes is the radiative recombination of an electron from the conduction band

**Table 8.1**  Properties and applications of LEDs and LDs

| Light emitting diode | Laser diode |
| --- | --- |
| Broad  beam, incoherent light | Narrow beam, coherent light |
| Simple to use | Requires control of current and temperature |
| Can be modulated up to 100 MHz | Can be modulated up to 10 GHz |
| Spectral width 30 to 100 nm | Spectral width < 5 nm |
| Optical power < 0 dBm | Optical power < 27 dBm |
| Economical | Expensive |
| Poor linearity | Good linearity |

with a gap from the valence band. The pre-requisite for a light source of sufficient intensity is the generation of a large number of recombinable electron-pairs. The physical properties of the boundary of p-doped or n-doped semiconductor material can be used specifically for this purpose.

The integration of external atoms of lower valence – of an "acceptor" – in the semiconductor crystal (e.g. elements from the third main group in a silicon lattice) creates an unsaturated condition that exhibits all the properties of a hole (p-doping). Conversely, the integration of a foreign atom of higher valence – a "donator" (e.g. arsenium in silicon) – creates an excess electron that stays in the conduction band even with a low thermal energy supply (n-doped material).

If the p-doped and n-doped semiconductor materials are brought into contact with one another, electrons and gaps can diffuse through the boundary. In the p-doped area, the diffused electrons fill up unsaturated combinations, causing an increased concentration of negative space charging close to the boundary. Conversely, the holes that diffuse into the n-doped material accept excess electrons, so that in the vicinity of the boundary a positive depletion region is created. The adjacent zones of positive and negative space charge form a potential barrier that prevents further diffusion of charge carriers through the boundary. As a result, a depletion of free charge carriers occurs in this zone (shown in Fig. 8.4 as depletion zone).

If an external positive voltage is applied to the p-doped area and a negative voltage to the n-doped area, then the height of the potential barrier and the width of the depletion zone are reduced. This enables electrons from the n-doped area to drift into the p-doped area or holes from the p-doped area to drift into the n-doped area.

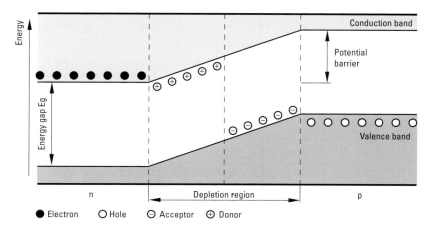

**Fig. 8.4** pn-transition of an LED without bias

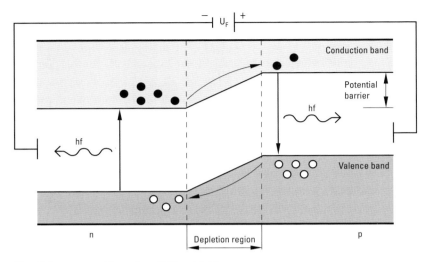

**Fig. 8.5** pn-transition of an LED with bias

In this way an increased number of recombinable electron-hole pairs is created by "injection" of minority charge carriers. In the semiconductor materials in which the radiative recombination has a higher probability, this "recombination current" can be used to implement a technical light source.

The radiation output of an LED is proportional to the current through the diodes. Where the currents are not too great, therefore, a good linearity can be observed between emitted radiant power and forward current.

When the temperature is raised, e.g. by self-heating of the diodes at higher control currents, the probability of a radiative recombination di-

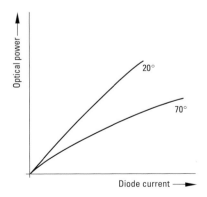

**Fig. 8.6** Characteristic of the LED at different temperatures

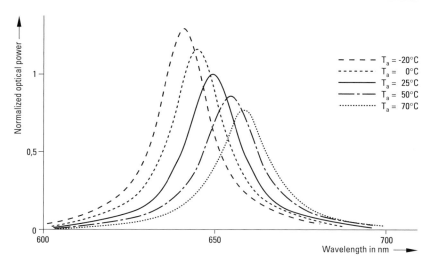

**Fig. 8.7** Temperature dependence of the spectrum of an LED

minishes. This is synonymous with a reduction of the emitted optical power. For this reason, deviations from the linearity can be recorded as the power increases (see Fig. 8.6).

A rising temperature not only causes the characteristic line to fall, but also a shift in the spectrum to higher wavelengths (see Fig. 8.7).

The wavelength of the emitted radiation depends on the size of the band gap, while the spectral width depends on the several influencing factors, such as the energy distribution of the charge carriers, the doping etc. Typical spectral widths are 25 to 40 nm FWHM (full width at half maximum) for LEDs with a 0.6–0.9 mm emission wavelength and 50–100 nm for emission wavelengths in the range of 1.1–1.7 μm.

The active area of an LED has typical dimensions of 300 μm × 300 μm and radiates non-directionally. Fig. 8.8 illustrates the radiation behavior.

The leap in the refractive index at the transition to the air, however, causes total reflection at relatively small emission angles (16.1° for $n = 3.6$). The power that is emitted from the active area at larger angles can no longer leave the semiconductor crystal through the crystal/air boundary, but under certain circumstances only after several reflections against the edges. This waveguides effect is exploited in the edge-emitting LEDs (see Fig. 8.9).

The typical dimensions of the active area is about 0.1 μm × 20 μm. The advantages of better launching characteristics of light from an edge-emitter into the fiber are counteracted by the disadvantages of a higher tem-

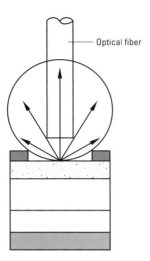

**Fig. 8.8**
Radiation characteristics of a surface-emitting LED (Lambert radiator)

perature dependence of the wavelength spectrum (and thus a shift to wavelengths of higher attenuation) in the optical fiber.

LEDs are easily modulated, as the optical output power depends on the diode current in an approximately linear relationship. The achievable modulation bandwidth is limited in principle by the lifetime of the minority charge carriers. The lifetime of the charge carriers in turn depends on the band structure of the semiconductors, the concentration of the doping, the number of injected minority charge carriers and the thickness of the active layer. LEDs with modulation bandwidths of 156 Mbit/s per second are commercially available.

To increase the coupling efficiency between LED and optical fiber, various methods are possible, for example the insertion of a spherical lens or Selfoc lens or a special development of the LED form.

**Fig. 8.9**   Radiation characteristics of an edge-emitting LED

## Laser diodes (LD)

While the emission of light in the LED is based on the spontaneous re-combination of electron-hole pairs, the laser principle additionally exploits the possibility of light amplification by stimulated emission (see Fig. 8.3). To facilitate amplification, the probability of an emission for the spectral range concerned must be greater than that of absorption. This is achieved by "pumping" the laser. The semiconductor is put into a particular condition, the "inversion state". The occupation number density in the upper energy level is then greater than that in the lower level. This "occupation inversion" can be achieved by an extreme doping of the n- or p-material. As in the case of the LED, the light emission is achieved by injecting minority charge carriers.

A laser refers to a light source with a tightly bundled beam, an almost monochromatic emission spectrum and a defined radiative phase behavior. An electromagnetic radiation field with such properties is called "coherent". This is achieved by selective optical feedback with the aid of an optical resonator that can be implemented in the form of two opposing mirrors. (Fabry-Perot resonator) (see Fig. 8.10).

By means of multiple reflections, waves in the resonator can be developed for particular discrete wavelengths. In the case of the semiconductor laser, the split end surfaces of the crystal act as a mirror, as about 30% of the radiation power is reflected from them due to the difference in the refractive index compared with air. Laser operation is then possible for those resonance frequencies of the resonator (longitudinal modes) for which the optical amplification exceeds the output coupling and absorption losses. (See Fig. 8.11).

The modes exhibit a competitive behavior with one another and fluctuate relative to time (mode noise). By carrying out specific measures it is possible to ensure that only a longitudinal mode remains under the amplification curve and is amplified. The laser then emits in single mode operation. By limiting the width of the resonator and the active zone, the oscillation of several lateral modes is avoided.

The ideal laser diodes exhibit a linear characteristic between the optical output power and laser current above a characteristic threshold $I_S$, at

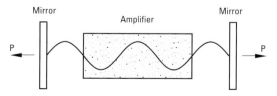

**Fig. 8.10**   Optical resonator

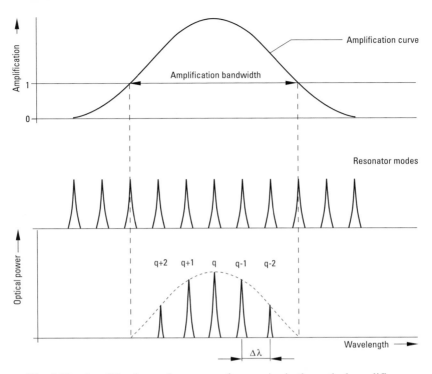

**Fig. 8.11** Amplification and resonance frequencies in the optical amplifier

which the laser operation is cut in (see Fig. 8.12). Below this threshold, the optical amplification is not sufficient. The light is emitted as in the case of the LED on the basis of spontaneous recombination.

The optical output power of the laser as a function of the injection current depends greatly on the temperature. The reason for this is the temperature dependence of the charge carrier concentration in the active layer as well as a probability of non-radiative recombination processes that increases as the temperature rises. As the temperature rises, the threshold current also increases and the slope of the characteristic line is reduced (see Fig. 8.13).

For the temperature dependence of the threshold current $I_S$, the following applies:

$$I_S\,(T + \Delta T) \approx I_S\,(T) \cdot \exp\,(\Delta T/T_0) \tag{8.3}$$

where $T_0$ is a material-specific characteristic temperature. The smaller $T_0$ is, the more sensitively the laser reacts to temperature changes. Typical

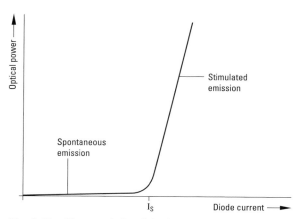

**Fig. 8.12**   Characteristic of the ideal laser diode

values [8.1] from the GaAlAs material system are: $T_0 = 120-230$ K; and for lasers from the InGaAsP material system: $T_0 = 60-80$ K. The close dependence on temperature demands monitoring and if necessary control of the semiconductor temperature or of the emitted radiance. To this end, the laser diode chip is mounted in the laser diode module which additionally contains a thermistor, a Peltier element and a monitor diode. With the aid of the Peltier element, the chip temperature is kept constant. The monitor diode detects the power emerging from the back laser mirror and keeps this constant with the aid of a control loop. As the laser

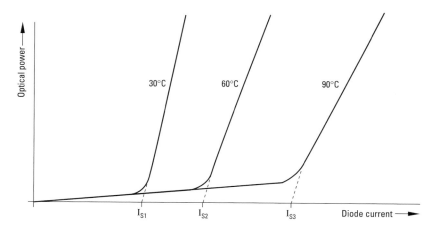

**Fig. 8.13**   Temperature dependence of the laser diode characteristic

diode ages, the laser current must be increased in order to keep the output power constant.

The temperature also affects the emission spectrum. As the temperature rises, the individual lines and the amplification profile are shifted to higher wavelengths:

$$\left.\frac{\delta\lambda}{\delta T}\right|_{profile} \approx \begin{array}{l} 0,24 \text{ nm/K for GaAlAs-Lasers} \\ 0,30 \text{ nm/K for InGaAsP-Lasers} \end{array} \qquad (8.4)$$

$$\left.\frac{\delta\lambda}{\delta T}\right|_{line} \approx \begin{array}{l} 0,12 \text{ nm/K for GaAlAs-Lasers} \\ 0,08 \text{ nm/K for InGaAsP-Lasers} \end{array}$$

As the shift varies in strength, it may result in mode jumps. In the case of the laser diode, as with the LED, the achievable bandwidth or the minimum rise time of the optical output power is also dependent on the mean life time of the minority charge carriers. Through the process of stimulated emission, the life, however, is considerably reduced, so that compared with the LED, considerably higher limit frequencies, up to about 10 GHz, are achieved.

## 8.2.2 Transmitters for use with plastic optical fibers

With the development of the plastic optical fiber, transmitter components are required that satisfy the special needs of this technology. For example, transmitter diodes have been, and are being developed, that emit at wavelengths that are adapted to the optical window (see Table 4.1). These transmitter diodes have small dimensions and offer broadband modulation.

### Light emitting diodes (LED)

In the first optical window for plastic optical fiber transmission (blue light), no suitable transmitters based on semiconductor are yet available for data transmission. For display applications blue LEDs are available.

*Green LEDs*

The lower attenuation of the plastic optical fiber in the second optical window (green light) in comparison with the third optical window (red light) makes this an attractive option, in particular for greater transmission distances. Then the lower power of these diodes is balanced by the lower attenuation. Conventional green GaP-LEDs achieve quantum efficiencies of only about 0.1%. That is less than a tenth of the quantum efficiencies of the red LEDs.

At the laboratory stage, a yellow-green LED was developed on the basis of the InGaAlP technology with a quantum efficiency of 0.7% at $\lambda = 573$ nm (8.2). Based on the familiar InGaAlP double heterostructure,

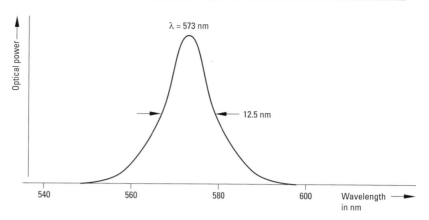

**Fig. 8.14** Emission spectrum of the green LED with a bias current of 20 mA

a series of further layers was added, creating a complicated structure. This LED has a considerably narrower spectrum, with FWHM = 12.5 nm, than previous diodes (see Fig. 8.14). This means that the spectrum of the diode both in terms of the peak wavelengths and the spectral width is almost ideally adapted to the attenuation minimum.

The attenuation minimum of the plastic optical fiber (see Fig. 4.2) at 568 nm is relatively broad, so that minor deviations of the peak wavelengths are not critical. By comparison, the peak wavelengths of the red LED must be exactly 650 nm, as the spectral attenuation minimum there is very narrow. What is more, the wavelengths in the manufacture of the red GaAlAs-LEDs is relatively difficult to control with the aid of the liquid-phase epitaxy procedure. The peak wavelength is usually shifted to greater wavelengths, where the attenuation of the plastic optical fiber is much higher. In addition, the peak wavelengths of the LED shifts at higher temperatures to higher wavelengths (about 7 nm shift for a 60 K temperature change). This makes the effectively measured attenuation in the red spectral region greater (see Table 4.3).

According to [8.3], the data rate of 10 Mbit/s NRZ (non return to zero) has been achieved with the diode specified above. The power launched into the plastic optical fiber was −17.5 dBm at $I$ = 100 mA. This corresponds to a mean output of −20.5 dBm. Assuming a receiver sensitivity of −32.5 dBm, the available budget is 12 dB. Taking into account a system reserve of 3 dB and a mean attenuation in the plastic optical fiber of 70 dB/km, a transmission length of more than 100 m is obtained. Further development of the green LED is quite possible in the near future, so that the mean input power will be −10 dBm at $I$ = 50 mA. Transmission lengths of more than 200 m will then be feasible.

*Red LED*

Despite the successes in the development of green LEDs and the lower attenuation of the plastic optical fiber in this wavelengths and, development work is concentrated in the red wavelengths. The reasons for this are the higher achievable data rates and the higher outputs. Typical parameters of commercially available red LEDs are listed in Table 8.2.

One interesting aspect is the use of specially selected LEDs ($\lambda \sim 660$ nm) which are able to function in a steady state as photodetectors in reverse bias operation and are thus particularly suitable for a "ping-pong" operation. Further developments are aimed in particular at increasing the output power, raising the coupling efficiency when launching into the plastic optical fiber and increasing the bandwidth.

Measures for increasing the coupling efficiency have been described in [8.4]. Although the plastic optical fiber has a high numerical aperture and a large core diameter, losses still occur when launching the light from an LED, as this radiates at a wider angle (Lambert radiator). A doubling of the coupling efficiency is to be equated with a doubling of the output power of the LED. In order to double the output power considerable technological effort is required.

For this reason any measures aimed at increasing the coupling efficiency are always worthwhile.

In order to solve the coupling problems, a reflector arrangement was developed which converts rays with an excessive angle of inclination to the optical access into rays with a lower angle of inclination, which are then within the numerical aperture of the plastic optical fiber (Fig. 8.15). With an arrangement of this type a doubling of the coupling efficiency has been achieved.

As a result of mode dispersion, a conventional plastic optical fiber (core diameter 0.98 mm, numerical aperture 0.5, step index profile) limits the bandwidth-distance product to about 45 MHz·100 m. To reduce the mode dispersion, the numerical aperture of the fiber of the plastic optical fiber is reduced. This, however, also requires a reduction of the numerical

**Table 8.2** Typical parameters of commercially available red LEDs

| Property | Technical data |
|---|---|
| Output power | $-3$ dBm |
| Peak wavelength | 650 nm ... 670 nm |
| Full width at half maximum | 20 nm ... 30 nm |
| Temperature coefficient of output power | $-0.02$ dB/K ... $-0.04$ dB/K |
| Temperature coefficient of peak wavelength | 0.12 nm/K |
| Temperature range | $0\,°C$ ... $70\,°C$ |

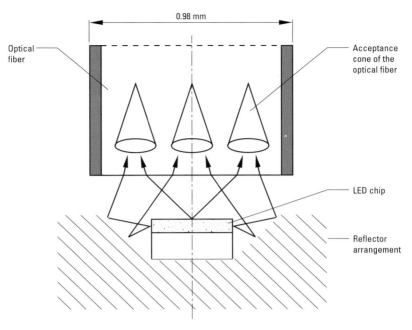

**Fig. 8.15** LED reflector arrangement

aperture of the source. An optimized structure was used in [8.5] to achieve a lower numerical aperture of the LED. Fig. 8.16 shows this annular LED with a double-ring electrode structure.

The annular upper contact is necessary in order to obtain an emission peak in the center of this structure. The light is only emitted by the surface of the LED. A molded plastic lens attached to the LED causes a reduction of radiative divergence from 70° to 10°.

**Laser diodes (LD)**

*Green LDs*

The green semiconductor laser diodes are not currently available commercially. Laser diodes that do not operate on the basis of semiconductors are not considered for communication transmission purposes due to their high price, greater volume and poor modularity. Their use is only conceivable for attenuation measurement of very great lengths of plastic optical fiber, for example in the production process. The available laser wavelengths, however, do not lie within the attenuation minimum of plastic optical fiber. These wavelengths are physically determined and cannot be changed. Table 8.3 lists some properties of green laser diodes.

89

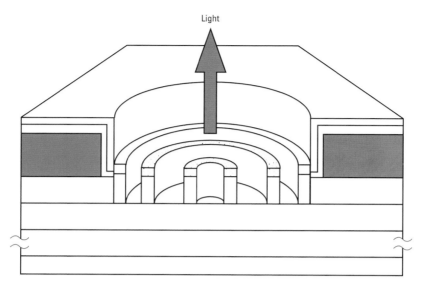

Light

**Fig. 8.16**   Annular LED (cross-section)

*Red LDs*

The typical parameters of commercially available red laser diodes are listed in Table 8.4. The higher performances relates to the higher wavelengths, so that – at least for greater transmission lengths – the increased fiber loss cannot be compensated in this range.

Further developments are concerned primarily with increasing performance and bandwidth: with the development of graded index plastic optical fibers the mode dispersion is sharply reduced, so that this is no longer the limiting factor for the bandwidth in the transmission system. For this reason it is advisable to increase the modulation speed of the laser. A report can be found in [8.6] on a high speed laser diode that operates at 4 GHz and 14 mW at approximately 650 nm. This is a modified

**Table 8.3**   Properties of green laser diodes

| Laser type | HeNe-Laser | Frequency doubled Nd:YAG-laser |
|---|---|---|
| Laser wavelength/nm | 543 | 532 |
| Power/dBm | $-3 \ldots 3$ | $5 \ldots 20$ |
| Fiber loss at above wavelength in dB/km | 119 | 92 |
| Price | high | extremely high for high performance |

**Table 8.4** Properties of available red laser diodes

| Property | Technical data |
|---|---|
| Output power | 5 dBm ... 27 dBm |
| Peak wavelength | 650 nm ... 690 nm |
| Full width at half maximum | < 1 nm |
| Threshold current | 35 mA ... 2000 mA |

version of the edge-emitting red laser diode that was developed for CD-ROM systems. This is an AlGaInP MQW laser (multiple quantum well) with a resonator length of 300 mm. The front/back resonator surfaces have been given a 30%/95% reflective coating. At a temperature of 25 °C a threshold current of 24 mA, a characteristic slope of 0.6 W/A and a working wavelength of 647 nm have been achieved. With the aid of a gradient index (GRIN) lens, light was launched into the plastic optical fiber achieving coupling efficiencies of approximately 25%.

The laser types discussed so far are edge-emitters: the resonator is arranged at the level of the semiconductor layers based on the laser (perpendicular to the current flow direction). The light leaves the semiconductor from a lateral surface. The advantages of this structure – in particular for laser operation – have been explained above.

Nevertheless, lasers have also been developed in which the radiation is launched perpendicularly to the layer structure (in the current flow direction) through one of the top surfaces. Lasers of this type are called vertical cavity surface emitting lasers (VCSELs). They require reflectors both above and below the active zone, so that a resonator can be formed perpendicular to the layer structure. These lasers were originally developed for transmission at 850 nm via glass optical fibers. A laser of this type was modified for the red wavelengths in [8.7]. The advantages of the VCSEL technology compared to LEDs are as follows:

▷ Low beam divergence (~ 10°) and circular beam profile, higher coupling efficiencies, no need for launching optics,

▷ Higher modulation rates (> 2.5 Gbit/s),

▷ Lower driver currents (typically 10 mA), lower power consumption in the driver circuit, reduction of heat dissipation in the module, improved reliability and temperature and behavior,

▷ Lower spectral widths (< 1 nm); thus achieving a lower fiber loss.

Currently, the VCSELs are operating at 670 nm and are not adapted to the loss minimum of the plastic optical fiber. At this wavelength the attenuation is about 300 dB/km. By comparison, the attenuation value at 650 nm is about 130 dB/km for plastic optical fibers. The current development work is concentrated on shifting the emission wavelength toward

91

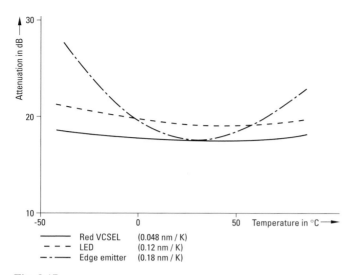

**Fig. 8.17**
Attenuation of 100 m of POF when measured with three different transmitters as
a function of temperature

650 nm and achieving the predicted low temperature dependence in practice as well. If one calculates the attenuation values that are obtained
when measuring 100 m of plastic optical fiber with three different transmitters, dependencies such as those shown in Fig. 8.17 are produced. The
loss minimum of the plastic optical fiber is 180 dB/km at 650 nm. Thus
the three transmitters operates at room temperature in the loss minimum.
The superiority of the VCSEL is clearly evident.

At present, VCSELs are suitable for short, but high bit-rate transmission
routes. For example, transmission over 13 m with a step index profile
plastic optical fiber has been implemented at 531 Mbit/s and with a
graded index plastic optical fiber at 1062.5 Mbit/s.

By means of a matrix arrangement of $2 \times 2$ VCSELs at a distance of
150 mm it was possible to launch four times the power into the plastic
optical fiber. With such an arrangement, the reliability of transmission is
increased by raising the redundancy.

# 8.3 Receivers

## 8.3.1 Fundamental properties

The receiver components must be characterized by:

▷ Great bandwidth,

▷ Compact and lightweight construction,

▷ Low noise,

▷ High sensitivity,

▷ High linearity.

The operating principle of the photo diode is based on the internal photoelectric effect which has already been explained in Section 8.1 and Fig. 8.2: Through the absorption of a photon with an energy that is greater than the energy of the semiconductor material, an electron-hole pair is produced. If it is possible to separate the pair by means of an electrical field before the electron and hole recombine, a measurable charge carrier concentration is created that is proportional to the number of photons. To separate the pairs in a photodiode, it is necessary to exploited the electrical field that is generated by space charges in the area of the depletion zone between p- and n-doped semiconductor material. This accelerates the released electrons toward the n-doped layer and the holes to the p-doped layer. This in results in an accumulation of positive charges in the valence band of the p-doped layer and negative charges in the conduction band of the n-doped layer. If both layers are connected to one another by a current circuit, then electrons flow from the n-doped to the p-doped layer (i. e. in the "depletion direction" of the diode), where they recombine with excess holes. Different semiconductor materials have different sensitivities to light of a specific wavelength. This sensitivity is called "spectral sensitivity" and specifies the generated currents strength relative to the optical power. The unit for this variable is an ampere/watt (A/W). Silicon is used mainly for wavelengths below 900 nm, as it is economical and available in large quantities. At higher wavelengths materials such as Germanium or InGaAs are used.

Fig. 8.18 shows the relative sensitivity as a function of the wavelengths for different detector materials.

Table 8.5 lists some absolute sensitivities that can be achieved today.

### Types of receiver element

A distinction is made between two types of receiver elements that are used for optical transmission:

▷ PIN photodiodes: this refers to a structure in which an i-zone (intrinsic) of non-doped semiconductor material is placed between the p- and

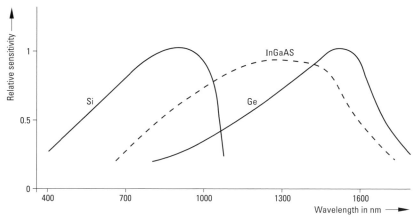

**Fig. 8.18**
Relative spectral sensitivity as a function of the wavelength of different detector materials

n-conducting semiconductor layers. This concept enables the absorption zone to be enlarged. The PIN photo diodes exhibits a very linear relationship between the incident beam and the generated photocurrent. It is suitable not only for analog, but also for digital applications. The required voltages are relatively low.

▷ Avalanche photodiode (APD): this permits an internal amplification of the photo current. From a single incoming photon, an avalanche of charge carriers is generated by means of impact ionization. Avalanche photo diodes have a very high sensitivity, but require a high, and well-stabilized operating voltage.

## 8.3.2 Receivers for transmission with plastic optical fibers

As can be seen from Fig. 8.18, only diodes on a silicon basis are considered for transmission with plastic optical fibers. The sensitivity drops to-

**Table 8.5**   Absolute sensitivity of receivers implemented today

|  | Silicon Si | Germanium Ge | InGaAs |
|---|---|---|---|
| $\lambda = $ 520 nm 1st optical window plastic OF | 0.25 | – | – |
| $\lambda = $ 570 nm 2nd optical window plastic OF | 0.3 | – | – |
| $\lambda = $ 650 nm 3rd optical window plastic OF | 0.4 | – | – |
| $\lambda = $ 850 nm 1st optical window glass OF | 0.55 | 0.3 | 0.2 |
| $\lambda = $ 1300 nm 2nd optical window glass POF | – | 0.65 | 0.9 |
| $\lambda = $ 1550 nm 3rd optical window glass POF | – | 0.9 | 0.95 |

**Table 8.6** Typical parameters of available photodiodes

| Properties | Technical data |
|---|---|
| Sensitivity | 0.18 A/W ... 0.7 A/W |
| Minimum detectable power at data rates <10 Mbit/s | −30 dBm |
| Rise time (10% ... 90%) | 1 ns ... 10 ns |

wards the lower wavelengths and is about 0.3 A/W in the green wavelength band ($\lambda \sim 568$ nm) and about 0.4 A/W in the red wavelength band ($\lambda \sim 650$ nm). In general, photo diodes that have been developed for optical transmission with glass optical fibers can also be used for plastic optical fibers, without having to carry out special material system developments as in the case of the transmitter elements. The receiver sensitivity is only $1-2.5$ dB less than at 850 nm. This is little in comparison with the relatively high attenuation values of the plastic optical fiber. Typical parameters of commercially available photo diodes are listed in Table 8.6.

The use of graded index plastic optical fibers necessitates not only high-bit-rate transmitters, but also broadband receivers. The diffusion of the charge carriers in a pn-structure is a very slow process. Charge carriers continue reaching the edge from the diffusion regions over a long period. Due to these stragglers, the photocurrent only decays slowly and the demodulation bandwidth is correspondingly small. Help is provided by a PIN structure: between the highly-doped p- and n-conducting area there is an intrinsic (i-conducting) zone. The abbreviation PIN thus describes the sequence of layers. PIN diodes are operated with negative bias voltage and with a load resistance.

A detector was described in [8.8], for example, that processes a 4 Gbit/s signal. This PIN photo diode with a 90 μm mesa-structure is aimed at reducing the effective surface of the pn-transition and thus lower in the depletion layer capacitance. The capacitance achieved at 10 V bias voltage, including the package capacitance, is 0.87 pF , which makes a processing speed in excess of 4 GHz possible.

In order to make the photodiode usable for transmission with a graded index plastic optical fiber, the 90 μm receiver diameter must be increased to several hundred μm. These specific demands on the size of the detector only exist in the case of transmission with a plastic optical fibers.

A receiver with a silicon PIN diode and an active area of 400 μm diameter has been described in [8.9]. The photodiode was coupled to the graded index plastic optical fiber with the aid of a GRIN lens. Using a PIN-FET receiver (PIN diode with subsequent field-effect transistor), a receiver sensitivity of $-16.9$ dBm was achieved with a bandwidth of 2.5 Gbit/s and a bit error rate of $10^{-9}$.

# 9 Connection systems

With the aid of suitable connection systems, the plastic optical fiber can be connected to the transmitter and receiver components described in Chapter 8.

## 9.1 Overview

The objective when developing the current connection systems for plastic optical fibers was to design the overall system as cost-effectively as possible. This involves both the housing for the transmitter and receiver elements and the connector, as well as the procedure with which the optical fiber is fixed in the connector and to how the fiber end faces are processed. Due to the large fiber core diameter, there is no need for the expensive precision components such as those used in glass fiber connection technology.

As a series of standardized and widely used connector systems were already familiar from glass fiber optic technology, these "standard systems" were adapted for plastic optical fiber technology in addition to the development of manufacturer-specific systems. The advantage of these "standard systems" is that the user is not dependent on one manufacturer.

In principle a distinction can be made between two different connection systems: the connectorless system and the connector system.

## 9.2 Connectorless systems

In these systems the plastic optical fiber core is inserted into the active components without a connector. This saves the cost of fitting the connector, as well is the connector itself. On the other hand, it is necessary to prepare the fiber end faces properly.

One example of such a clamp connector is the SFH series from Siemens, in which the plastic optical fiber cores is fixed to the active components by screwing down a removable coupling nut (Fig. 9.1). The particular advantage of this system is that the 0.6 mm thick buffer around the fiber does not have to be removed. The entire component is made of plastic and is suitable for direct mounting on a circuit board. The fiber end face is processed either by cutting or grinding and polishing.

**Fig. 9.1**   SFH series housing

**Fig. 9.2**
Ferrule for hot-plate
connectorizing

In order to use hot plate technology (see section 10.4) for the preparation
of the end faces, a suitable ferrule (see Fig. 9.2) can be used.

## 9.3  Connector systems

Connector systems are characterized by a connector that is fitted to the
end of the plastic optical fiber. This is inserted as an easily removable
connection into the corresponding receptacle of the active component.
Various connector systems are illustrated in Figs. 9.3 through 9.6.

Table 9.1 lists the connector systems that are currently most widely used.
Simplex connectors, i.e. for connecting a single fiber core, are available
for all the systems described. In addition, duplex versions are available in
some of the systems (F07, Versatile Link). These are designed for the
connection of two cores in one connector body, which is a particular advantage for the frequently used bi-directional data transmission. In principle, the procedures mentioned in Chapter 10 for processing the end
faces are possible for all connector systems.

Connectors and plastic optical fiber cores are joined either by crimping
or by clamping the core to the connector. The crimping method requires

**Table 9.1**  Properties of the most widely used connector systems

| System | Standard | POF connection method | Connection method between connector element and active components | Connector with receptacle for strength member |
|---|---|---|---|---|
| FSMA | consistent with IEC 874-2 | Clamping or crimping | Screw | available |
| BFOC | consistent with IEC 874-10 | crimping | Bayonet | available |
| Versatile Link[1] | manufacturer-specific | Clamping or crimping | Snap in | not available at present |
| F05 | JIS C 5974 | Clamping | Snap in | not available at present |
| F07 | JIS C 5976 | Clamping | Snap in | not available at present |

[1] Trademark of Hewlett Packard

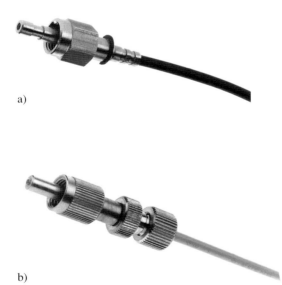

a)

b)

**Fig. 9.3**
FSMA connector: a) FSMA connector; crimped plastic optical fiber; b) FSMA connector; plastic optical fiber clamped

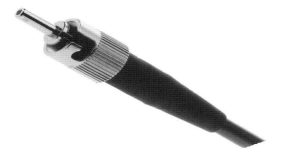

**Fig. 9.4** BFOC connector; crimped plastic optical fiber

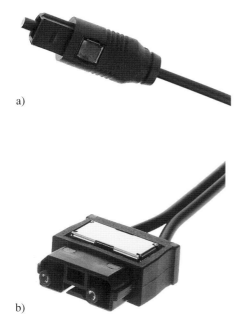

a)

b)

**Fig. 9.5**
a) F05 simplex connector; plastic optical fiber clamped;
b) F07 duplex connector; plastic optical fiber clamped

a special crimping tool in every case. No special tool is required for the connectors that hold the fiber by means of clamps, which is a particular advantage when connectorizing in the field.

A cable strength member on the connector is often an advantage in industrial applications. Practical experience has shown that it is usually sufficient to use a fiber optic cable with strength members without attaching the strength member to the connector. The greatest loads caused by tensile forces evidently occur during installation. After installation, the tensile strength of the simple fiber core of about 5 N is sufficient, which means that the strength member on a special connector is not necessary. The decision about this, however, must be made in each parti-

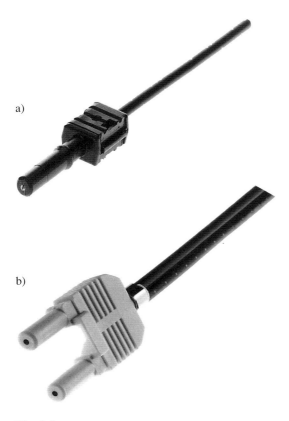

**Fig. 9.6**
a) Versatile Link simplex connector; plastic optical fiber clamped;
b) Versatile Link duplex connector; plastic optical fiber crimped

cular case. An overview of connectors currently available in which the strength member of the cable can be attached to the connector is shown in table 9.1

At the start of the development, industrial users were very skeptical towards plastic connectors. This initial skepticism however has since approved unfounded due to the positive results in practice.

## 9.4 Adapters

Adapters are passive components for the connection of two optical fibers. They are used on one hand for repairs, but can on the other hand be deliberately integrated into a transmission system, if the fiber optic link is to be easily separated. Adapters are available for most of the systems listed in section 9.2 and 9.3.

## 9.5 Special designs

### Active connectors

Active connectors are based on the idea that only electrical contacts are used externally. The active components for electro-optical conversion, as well as the coupling point of the fibers to these, are thus contained within the connector. These systems are particularly robust, as the danger of dirt getting onto the end faces is considerably restricted. In addition, they are quickly accepted by the user, as the technology using external electrical contacts is already familiar.

### Hybrid connectors

Hybrid connectors have electrical contacts in addition to the optical contacts. Connectors of this type are usually deployed where an operating voltage has to be carried in addition to the data transmission. This method does, however, lose the benefit of electrical isolation offered by the optical fiber. Due to their compact design, however, these connectors prove very useful where electrical isolation is not an essential requirement. The hybrid connectors are available in a wide variety of designs, both as passive and as active connectors.

On the basis of connector systems that are familiar from electrical connection technology, a host of connector systems have been developed for plastic optical fibers. An example of such a system is the sub-D connector (see Fig. 9.7). With the system illustrated two plastic optical fibers and two copper contacts can be connected. The active components are installed in the socket that can be mounted on a printed circuit board.

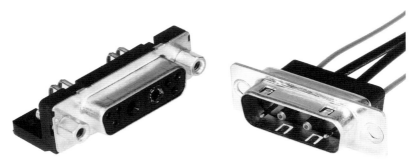

**Fig. 9.7**
Hybrid connector designed as sub-D connector for plastic optical fiber (works photo by Harting KG)

**Fig. 9.8**
Hybrid connectors for 2 plastic optical fibers and 4 copper conductors (works photo by Harting KG)

Fig. 9.8 shows a further example of an extremely compact hybrid connector. With this system, two plastic optical fibers and four copper conductors can be connected simultaneously. At the connector, the hybrid cable is removed from the connector frame via a PG screw. The active components for the electro-optical conversion of the signal are integrated on the socket side. These are fitted by the manufacturer according to the wishes of individual customers. The electrical contacts can be withstand a load of up to 10 A. The entire connector is splash-proof in compliance with IP 65 and can be assembled by the user.

# 10 End-face processing procedures

Following the description in the preceding chapter of connectors for connecting a plastic optical fiber into the active components, we will now consider the transition point at the end of the fiber: the fiber end face.

## 10.1 Principles

The quality of the end face of the plastic optical fiber has a crucial bearing on the quality of the fiber-to-fiber, transmitter-to-fiber or fiber-to-receiver coupling. It is characterized by the parameters of attenuation and reflection.

Minimum loss, achieved though the quality of the end face, is not always first priority. In many cases, several dB of loss are tolerated, especially if economic – i.e. time-saving – connectorization is required. Thus there are currently several methods for end face processing:

▷ Cutting

▷ Grinding and polishing

▷ Melting (hot plate method)

▷ Cutting with a laser

▷ Cutting with a hot blade

▷ Machining methods

The first three procedures are explained below. The last three procedures listed have not acquired any great significance so far, although machining methods in the form of sawing or cutting are perfectly suitable for mass production.

Losses at coupling points, caused by end face processing, can occur at the end face, particularly as a result of fiber end face separation, surface roughness and reflection losses.

### Fiber end face separation

If the gap $s$ between the end faces of the fibers is greater than zero, the emitted power can no longer be transmitted fully into the connected optical fiber (see Fig. 10.1). The emitted radiation cone (solid line) is only partly accepted (dotted line) by the connected optical fiber.

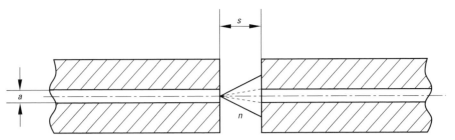

**Fig. 10.1**   Coupling loss due to fiber end face separation

The coupling loss (attenuation) is calculated from:

$$A = -10 \log \left( 1 - \frac{s \cdot NA}{3 \cdot n \cdot a} \right) \tag{10.1}$$

This formula applies to uniform mode distribution and step index profile. In this formula, $s$ is the separation distance, $NA$ is the numerical aperture, $n$ the refractive index in the intermediate gap and $a$ is the core radius.

**Fresnel reflection**

If the end face is exposed to air, Fresnel reflection occurs. By approximation, a perpendicular incidence can be assumed. Then the reflected proportion $R$ (dimensionless variable) is calculated from:

$$R = \left( \frac{n_1 - n_2}{n_1 + n_2} \right)^2 \tag{10.2}$$

where $n_1$ is the refractive index of the fiber core material and $n_2$ the refractive index of air.

Example:

Refractive index of the plastic optical fiber (PMMA) $n_1 = 1.492$
Refractive index of the air $n_2 = 1$
$R = 0.039 \equiv 14.1$ dB

For the transmitted signal $T$, the following is obtained:

$$T = 1 - R \equiv 0.96 \cong 0.17 \text{ dB} \tag{10.3}$$

In other words, as a result of an end face reflection on an ideal surface, the signal suffers a loss of 0.17 dB. At a coupling point between two plastic optical fibers, this loss is then 0.35 dB. This loss can only be avoided, if a medium with an adapted refractive index is inserted between the end faces (immersion). To this end, the use of immersion gel among other things has been examined in [10.1]. The data specified applies for a wavelength of 660 nm. This also applies for all subsequent results in this chapter.

**Surface roughness**

The problem of surface roughness exists in all the methods discussed below, in which the quality of the tool used for this purpose plays an important role. A rough surface creates scattering centers which reflect the light both forward and backward.

Scattering in the forward direction means that the light rays change their direction and under certain circumstances leaves the acceptance range of the optical fiber, resulting in losses.

Scattering in the backward direction increases the proportion of the reflected signal, which becomes measurable in particular if the Fresnel reflection at the end face has been suppressed (for example, as a result of refractive index adaptation). The size of the reflected signal depends on the roughness. The greater the grain, the greater the roughness depth and thus the reflection and the loss in transmission. For example, the loss due to reflection can be reduced by about 0.5 dB by reducing the grain size from 30 μm to 0.3 μm.

## 10.2 End face processing by cutting

Cutting is the simplest method but produces the lowest quality. The plastic optical fiber (PMMA) is relatively hard which makes it difficult to cut this material perfectly.

A great deal is demanded of the cutting blades in terms of material, thickness, cutting angle and surface roughness. The material must have a high endurance and be resistant to corrosion. A blade ground on just one side has proved particularly practical. It is especially important to replace the blade regularly, as it becomes blunt very quickly through cutting the relatively hard PMMA.

This method therefore results in high insertion loss. For this reason, and because the results are difficult to reproduce, this method is seldom used. It is used mostly for connectorizing short cables. An immersion gel, which is applied between the connector end face and the cut fiber, can improve the reproducibility of this method. In particular, this opens up possible new applications for this method in mass production (see also [10.1]).

## 10.3 End face processing by grinding and polishing

This is a well-developed procedure and familiar from glass optical fiber technology. The procedure is as follows:

▷ Removal of the plastic optical buffered fibre

▷ Inserting and fixing the plastic optical fiber in the connector

▷ Cutting off the plastic optical fiber so that it projects by a few tenths of a millimeter beyond the connector end face

▷ Grinding the plastic optical fiber flat using 600 grain abrasive paper on a hard and smooth based with the aid of a suitable grinding disc

▷ Polishing the plastic optical fiber end face using polishing papers with decreasing grain size (12 µm, 1 µm, 0.3 µm)

The grinding and polishing procedure is performed by moving the grinding disc in a figure of eight (see Fig. 10.2).

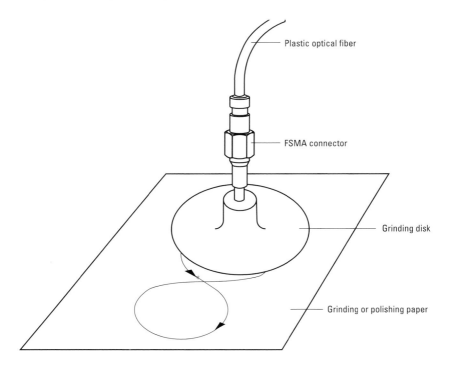

**Fig. 10.2**
End-face processing by grinding and polishing using the example of an FSMA connector

Grinding discs are commercially available today for most popular types of connector. They guarantee the exact vertical positioning of the fiber end face on the abrasive or polishing paper. It should be noted, however, that the grinding disc itself is worn down by pressing onto the abrasive paper. Especially in series production, regular replacement of the grinding disc is essential.

The typical attenuation values specified by several manufacturers are in the range between 0.6 dB and 1.6 dB. The lower figure is a very good result as a Fresnel reflection loss of 0.35 dB is included in the specified attenuation value. The procedure is applied above all in the installation of plastic optical fibers in the field.

## 10.4 End face processing by melting (hot plate procedure)

The hot plate procedure was specially developed for plastic optical fibers. Its particular advantages are its speed and the high rate of reproducibility (see Fig. 10.3).

The basic concept is based on the fact that the PMMA material of the plastic optical fiber can be deformed at temperatures of about 160 °C. During this plastic deformation the end face is pressed against a hot, highly polished surface. Using a special combination of heating and cooling a very smooth end face of the plastic optical fiber is achieved. The procedure is as follows:

▷ Removal of the plastic optical buffered fibre

▷ Inserting the plastic optical fiber into the connector and fixing it (by means of adhesive, crimping or screw clamp)

▷ Cutting the plastic optical fiber so that it projects by about 0.5 mm beyond the connector end face (depending on connector type used)

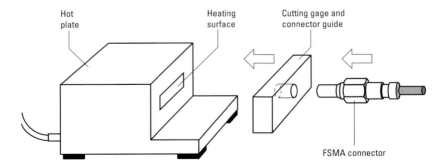

**Fig. 10.3** End face processing with a hot plate

107

▷ With the aid of the guide channel, the connector is pressed at right angles onto a hot and highly polished surface (approximately 160 °C). The surplus material flows into a sink in the connector end face, so that the plastic optical fiber terminates in alignment with the connector end face after melting.

▷ The polished surface, and with it the heated PMMA, is cooled down.

▷ The finished connector is now removed from the polished surface.

Depending on the type of hot plate used, the entire process of melting and cooling takes between 5 and 30 seconds. The hot plate is heated and cooled automatically. Fig. 10.4 shows a hot plate with power supply unit that is simple to operate and achieves very good reproducible results. A connector guide simplify the perpendicular positioning of the connector.

The disadvantage of this procedure is that the optical fiber core is widened in the vicinity of the connector end face. The typical value for the depth of a sink is 0.2 mm (see Fig. 10.5). This means that the plastic optical fiber cores in a connector-connector coupling with the original diameter of 0.98 mm, for example, have a gap of 0.4 mm. The fiber end face separation causes a loss of emitted power.

**Fig. 10.4**  Hot plate with power supply unit

**Fig. 10.5**  Sink in the FSMA connector

This loss can be calculated using equation 10.1, in which $s = 0.4$ mm, $NA = 0.47$, $n = 1.492$ and $a = 0.49$ mm, and attenuation of 0.39 dB is produced.

It is apparent that this loss is relatively small. Combined with the unavoidable Fresnel reflection loss, the minimum loss for a fiber-to-fiber coupling is 0.74 dB. The sink has virtually no effect on the coupling between LED and optical fiber, or between optical fiber and receiver.

Various designs of sink have been tried out [10.2]. In the sink, the deformed material forms a thin layer, giving rise to possible losses through axial offset. These can be minimized by the use of a V-shaped sink (Fig. 10.6). Typical attenuation values that have been specified by several manufacturers lie within the range from 0.8 to 1.6 dB, although the values differ widely. It should be remembered that the losses decrease with a smaller numerical aperture.

In this connector the light is guided to the connector end face within the unchanged optical fiber core, whereby no additional losses occur.

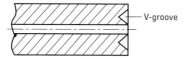

**Fig. 10.6**   V-groove in the hot-plate connector

**Fig. 10.7**   Wire stripper for plastic optical fibers

Finally, it is essential to note one fundamental point that applies to all the procedures mentioned. When removing the 0.6 mm thick protective buffer, it is essential to use a tool, such as the one illustrated in Fig. 10.7, that is guaranteed not to damage or scratch the fiber. The use of unsuitable tools can have a negative effect on the insertion loss. The diameter of the cutter opening is 1.3 mm.

## 10.5  Comparison of procedures

The individual procedures are compared with one another in table 10.1.

**Table 10.1**
Comparison of typical insertion loss and scattering of the individual procedures

| Procedure | Insertion loss | Scattering |
|---|---|---|
| Cutting | 2–3 dB | approx. 15% |
| Grinding and polishing | 0.6–1.6 dB | approx. 5% |
| Hot plate | 0.8–1.6 dB | approx. 2% |

# 11 Passive optical components

Optical attenuators and couplers are defined as passive optical components. Attenuators are used, for example, when the optical signal is too strong, in order to avoid overdriving the receiver. Couplers are essential in the construction of optical networks, for example in various methods of transport (cars, aircraft, trains), in LANs (for short distances or in-house applications), in industrial automation (machine tools, robots) or for sensor applications. They are used for distributing optical signals onto several signal paths, or merging these signals from several signal paths.

## 11.1 Couplers

### 11.1.1 Basic properties

The following demands are made on the couplers:

▷ Low loss

▷ Easy handling

▷ Reproducible coupling behavior

▷ Lower manufacturing costs

▷ Small dimensions

▷ Thermal and mechanical stability

▷ Low mode dependence

▷ Good isolation between the inputs

The coupler designs vary greatly according to the number of inputs and outputs. The designation of a coupler contains this information as follows: number of inputs × number of outputs, for example

▷ $1 \times 2$: Y-coupler

▷ $2 \times 2$: X-coupler

▷ $1 \times N$ or $N \times N$: Star coupler

Regardless of these diverse configurations, the definition of the major coupling parameters can be reduced to a simple X-structure (see Fig. 11.1).

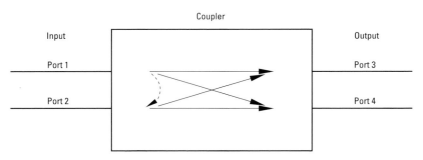

**Fig. 11.1** Principle of a 2×2 coupler

The following definitions then apply (in which light powers in μW are used):

▷ Excess loss

The excess loss (*EL*) in dB is a measure of the losses caused by the coupler, i.e. ultimately a measure of quality. It refers to the total power coupled out at the ports 3 and 4, $P_3$ and $P_4$, in relation to the input power $P_1$ at port 1:

$$EL = -10 \log \frac{P_1}{P_3 + P_4} \tag{11.1}$$

If there are no losses, then *EL* = 0 dB.

▷ Insertion loss

The insertion loss (*IL*) in dB characterizes the decay of the light power from one input to a single output:

$$IL = -10 \log \frac{P_3}{P_1} \text{ or } -10 \log \frac{P_4}{P_1} \tag{11.2}$$

It depends on the one hand on the coupling ratio, but is also influenced by the excess loss. In the case of the symmetrical division, *IL* = 3 dB, if there are no losses (excess losses).

The coupling ratio (*CR*) describes the ratio of the power at port 4 to the power at port 3.

$$CR = \frac{P_4}{P_3} \tag{11.3}$$

Thus it specifies how the power is divided between the two output ports. In the case of the symmetrical division, *CR* = 1.

▷ Directivity

The directivity ($D$) s given in dB and is a measure for the ratio between the emitted light power from an unswitched input and the launched power. Input and outputs are located on the same side of the coupler.

$$D = -10 \log \frac{P_2}{P_1} \tag{11.4}$$

▷ Uniformity

The uniformity ($U$) in dB is of importance primarily for multi-port couplers and specifies the difference between the insertion losses of the worst and best ports.

$$U = IL_{\max} - IL_{\min} \tag{11.5}$$

Example:

For the configurations shown in Fig. 11.2, the following parameters are known:

Output power of the transmitter: 0 dB/m
Attenuation of the optical fiber  0.16 dB/m
Power at receiver 1:              −8 dBm
Power at receiver 2:              −9 dBm

Taking into account the attenuation of the preceding plastic optical fiber (10 m), the power at the coupler input is as follows: $P_1 = -1.6$ dBm

Taking into account the attenuation of the subsequent optical fiber (20 m), the power values directly at the coupler outputs are as follows: $P_3 = -4.8$ dBm, $P_4 = -5.8$ dBm

From equations 11.1 through 11.3 and 11.5, the following coupler parameters can be calculated:

Excess loss:      $PL = 0.66$ dB
Insertion losses: $IL = 3.2$ dB or 4.2 dB
Coupling ratio:   $CR = 0.79$
Uniformity:       $U = 1$ dB

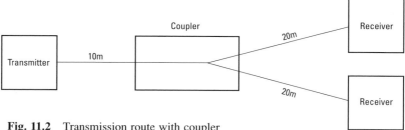

**Fig. 11.2**  Transmission route with coupler

## 11.1.2 Types of coupler

### Butt couplers

The butt coupler is the simplest solution for a $1 \times 2$ coupler. The polished optical fiber end faces are positioned together as shown in Fig. 11.3 and joined with an index-matching resin. The advantages of this method are, on the one hand, its simple manufacture and, on the other hand, the fact that the coupling ratio can easily be adjusted by shifting the launching fiber to one side.

The disadvantage is the launching of the light outside the axis and the high excess loss as a result of the poor surface coverage. This is about 1.1 dB for a coupling ratio of 50%.

*Fork-type coupler*

In the fork-type coupler the output fibers are ground in such a way that the end faces completely cover each other (see Fig. 11.4).

The disadvantage of the fork-type coupler is the costly grinding and difficult alignment, which means that only $1 \times 2$ or $2 \times 2$ configurations are possible at a reasonable cost.

Extensive investigations on the basis of these configurations have been published in [11.1]. Typical values obtained on Y-couplers (core diameter 1 mm, numerical aperture 0.5) were as follows: $EL = 2.5 \pm 0.24$ dB, $IL = 5.52 \pm 0.32$ dB, $D = 16.8 \pm 0.8$ dB and $CR = 1.08 \pm 0.04$ dB

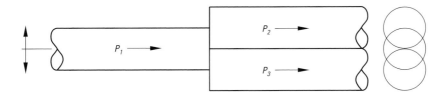

**Fig. 11.3**  Butt coupler

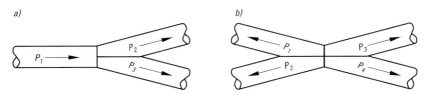

**Fig. 11.4**  Fork-type coupler  (a) 3 port  (b) 4 port

### Core fusion couplers

The core fusion coupler in its simplest form is a $2 \times 2$ coupler, although in principle any $N \times N$ configurations can be implemented. The optical fibers are brought into contact with each other over a specific length and fused together (see Fig. 11.5). Plastic optical fiber is fused by means of hot air or ultrasonic welding [11.2]. The advantage of this procedure is the ease of alignment and the fact that no expensive grinding and polishing process is necessary. The disadvantage is that expensive equipment is required.

In [11.2] two 1 mm PMMA plastic optical fibers were melted together using the ultrasonic welding procedure without removing the cladding. The coupling ratio achieved in this process depends on the fusion zone and on the diameter of the plastic optical fiber. Nevertheless, a certain coupling ratio cannot be exceeded. Only about 43% of the launched light power could be transferred into the fused optical fiber. This means that the coupler is always asymmetrical. The following typical values were achieved: $EL = 0.7$ dB, $IL = 3.3$ dB or 4.1 dB.

### Taper coupler

The taper coupler is constructive to fibers that are tapered in the coupling region. This principle is used with great success in the single mode melt couplers. The light power that is to be coupled into tea fused fiber, is unlimited.

### The bend coupler

In a bend coupler the input optical fiber is bent around a radius $r$ (see Fig. 11.6). Higher order modes leaves the fiber at the bend and travel through a transparent resin into a connected fiber with a smooth end face.

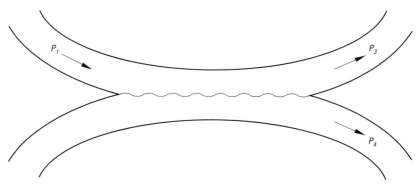

**Fig. 11.5**  Core fusion coupler

115

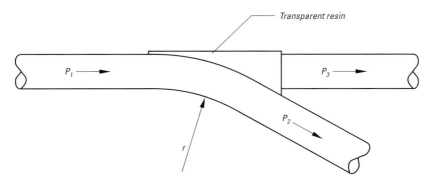

**Fig. 11.6**   Bend coupler

The coupling ratio, but also the insertion loss, increases as the bending radius is reduced. This means that only about 35% of the launched light power can escape from the bend without giving rise to excess losses [11.8].

If the input optical fiber is bent around 180°, two optical fibers can be coupled using the method described above, creating a 4-port coupler.

The advantages of the bend coupler are its uncomplicated assembly and the possibility of setting a specific coupling ratio in the simplest possible way. The disadvantage is its dependence on mode distribution and the limited coupling ratio.

## Core facet couplers

The core facet coupler is very similar to the core fusion coupler, but instead of fusing the optical fiber cores together over a certain length, they are mechanically processed and optically linked using an immersion gel.

The advantage of this procedure is that a defined geometry is achieved in the coupling region and it is thus more easily reproducible. The disadvantage is the expensive grinding and polishing as well as the problems of alignment.

No core facet couplers are yet known for plastic optical fibers.

## Couplers with the mixer elements

Couplers with mixer elements are star couplers. They can be constructed as a transmission type with fixed inputs and outputs wise a reflection type in which each optical fiber can be both and input and outputs. In the reflection type, the end face is reflected internally.

The elements most commonly used as mixers are mixer chips or mixer rods. The mixer rod is a smooth piece of waveguide, whose thickness corresponds to the diameter of the coupled optical fiber. The diameter of the mixer rod is dimensioned in such a way that all optical fibers can be connected to the end faces of both sides. By means of the mixer, the light from the input optical fiber is distributed more or less evenly among the output optical fibers. This requires a specific length of the mixer element.

The disadvantage is the fact that there are blind spots between the individual optical fibers into which no light can be launched. Depending on the configuration, this creates an excess loss in the order of 1 dB.

The advantage is that star couplers can be implemented with any number of inputs and outputs. The alignment of the components, however, is very time-consuming and therefore expensive.

Fig. 11.7 shows a schematic representation of a $7 \times 7$ star coupler with mixer chips (transmission type). According to [11.8], a component of this type has been manufactured with the following parameters: $EL = 2.68\,\mathrm{dB}$ to $3.44\,\mathrm{dB}$, $IL = 10.54$ to $12.83\,\mathrm{dB}$.

Various star couplers are described in [11.3] which have been constructed with the aid of a mixer rod. These mixer rods are transparent and have internally reflecting surfaces. They exhibit different geometries: right angle, cylindrical, conical and tubular. Couplers with $6 \times 6$, $7 \times 7$ and $16 \times 16$ have been assembled. In the last of these a mean insertion loss of 15.2 dB and a uniformity of 1.7 dB was achieved. The theoretical loss for a split of $1:16$ is 12.0 dB. This means that the excess loss is 3.2 dB.

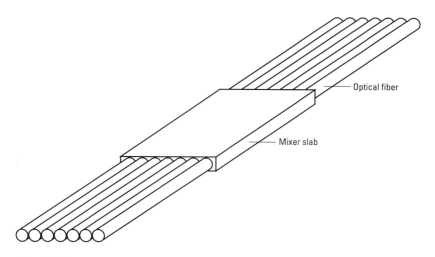

**Fig. 11.7**  $7 \times 7$ star coupler with mixer chips

117

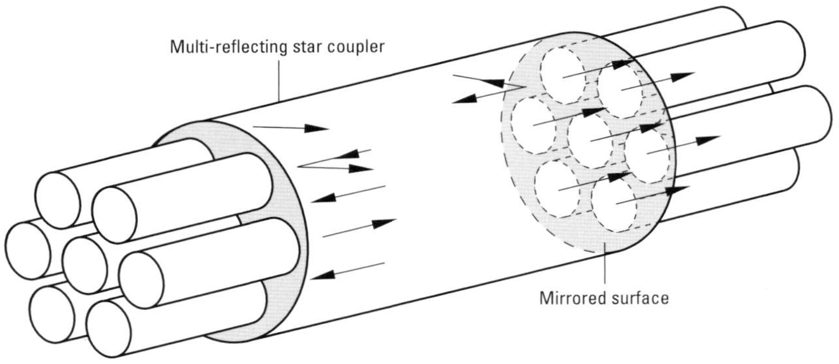

**Fig. 11.8** Multiple reflecting star coupler with mixer rod

In [11.4] an internally gilded tube was used as a mixer rod, the core area of which contained quartz. On bending the rod, the influence of the mixer lengths and the bending radius on the uniformity of the power distribution was examined.

This was carried out not only with regard to the output optical fiber, but also with regard to the mode distribution.

In [11.5] an optical fiber was coupled centrally to the input side of the mixer rod in a transmissive $1 \times 7$ star coupler. On the output side, seven tightly packed optical fibers were connected.

Another interesting proposal is a multiple reflecting star coupler (see Fig. 11.8). This does not select the densest packing of the optical fibers, but instead larger intermediate gaps are deliberately left which are mirrored internally. This permits a coupling between all, even adjacent, optical fibers. The disadvantage is the high excess loss of this arrangement.

**Couplers based on waveguides**

Branching structures based on a level waveguide can be constructed with the aid of the LIGA procedure (lithography, galvanic forming, deformation) [11.6, 11.7], in which both positive and negative structures are possible (see Fig. 11.9).

In this procedure the main loss occurs at the transition from the quadratic cross-section of the waveguide to the circular cross-section of the optical fiber. In this respect a $1 \times N$ coupler with a large $N$ is considerably more efficient, because the coupling loss due to change of cross-section occurs only twice, i.e. at the end face and at the rear surface, as in a simple $1 \times 2$ structure. In addition, the manufacturing costs do not rise with the number of ports, but if anything they are degressive.

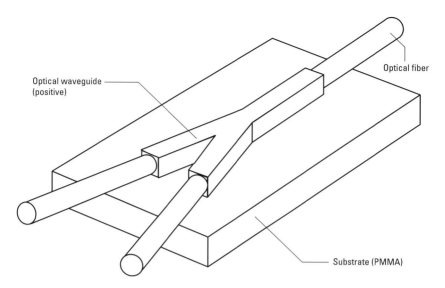

**Fig. 11.9**   Coupler based on an optical waveguide (positive structure)

The best results have been achieved with negative structures. The following data refers to a plastic optical fiber with a 1 mm diameter and a numerical aperture of 0.5:

$1 \times 2$ coupler: $IL = 4.7$ dB, $EL = 1.7$ dB,

$1 \times 4$ coupler: $IL = 8.9$ dB, $EL = 2.9$ dB.

### 11.1.3  Coupler topologies

Using the designs described above, all known topologies can be manufactured: bus, star and ring. Thus it is also possible, by cascading $1 \times 2$ couplers, to construct a $1 \times N$ bus structure, and by connecting $1 \times 2$ couplers to construct an $N \times N$ star structure. This is demonstrated in Fig. 11.10 on the basis of an $8 \times 8$ structure.

A star coupler can serve as a star point to connect several transmitters and receivers of equal status with one another (see Fig 11.11).

If two Y-couplers are not connected to form an X, but instead to form a W, a ring structure can be created (see Fig 11.12).

119

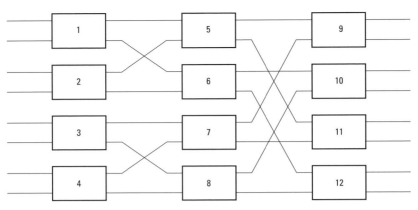

1 to 12: 2 x 2 coupler

**Fig. 11.10** Configuration of an $8 \times 8$ star coupler comprising 12 $2 \times 2$ couplers

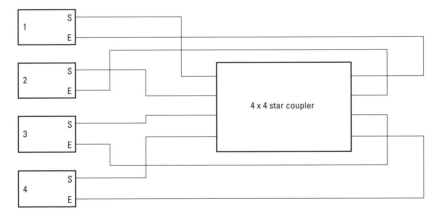

**Fig. 11.11** System structure with $4 \times 4$ star point

## 11.2 Attenuators

When using attenuators and couplers it should always be guaranteed that the mode distribution remains unchanged as far as possible. An interference in the mode distribution causes an undefined attenuation of the component. For example, a different attenuation is measured immediately after the component than immediately after a corresponding subsequent optical fiber (even if the attenuation of this subsequent optical fiber

120

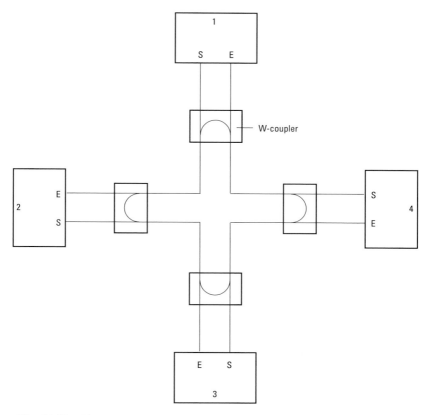

**Fig. 11.12**  Ring structure with W-couplers

is mathematically eliminated), because the mode conversion processes in the subsequent optical fiber causes the original mode distribution to be gradually restored, leading to additional losses.

In [11.8] an attenuator with a low mode dependence was proposed: a substrate is coated with a thin layer of PMMA material in which carbon powder is distributed. Together with the end face of a plastic optical fiber, this film is pressed against a hot plate. The thin layer melts into the end face of the plastic optical fiber and is detached from the substrate.

This creates a thin neutral step filter on the optical fiber end face, in which the loss depends on the thickness of the carbon in the PMMA. A filter of this type only has a very low wavelength dependence. This is an economical and compact arrangement that can be implemented without the use of lenses.

# 12 Using polymer clad fibers for longer distances

Sometimes it is necessary to span distances that exceed the current possibilities of plastic optical fibers. It is, of course, possible to build such systems with plastic optical fiber by using optical repeaters, for example, or by optimizing the transmitters/receivers. Another elegant solution is offered by the use of polymer clad fiber (PCF). This special type of fiber is being used on an increasing scale, especially in the industrial environment. What makes PCF so interesting? To sum it: PCFs can be connected to the same hardware as specified for plastic optical fiber and due to their special construction are exceedingly rugged and easy to connectorize. The sections below describe the important properties of this fiber.

## 12.1 PCF construction

The PCF consists of a glass core and a polymer-based cladding (see Fig. 12.1), which means that it is neither pure glass nor a pure plastic fiber, but a hybrid. This fiber is available in various diameters. For the application in question, the fiber with a core diameter of 200 µm and a cladding diameter of 230 µm has been widely accepted.

The fiber code is explained below, consistent with DIN VDE 0888, Part 4, using the example of a 230 µm fiber:

*Code*

F – K200/230 10 A 17
F – Fiber
K – PCF with step index profile
200/230 – Core/cladding diameter
10 – Attenuation coefficient in dB/km
A – Wavelength 650 nm
17 – Bandwidth-distance product 17 MHz · 1 km

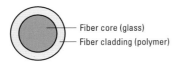
— Fiber core (glass)
— Fiber cladding (polymer)

**Fig. 12.1**   Construction of a PCF with step index profile

Due to its refractive index, the polymer used for the cladding determines the numerical aperture of the fiber on the one hand and the mechanical properties of the fiber on the other hand. PCF is extremely easy to use. And thanks to its large cross-section, this fiber only breaks at very tight bending radii.

## 12.2 Properties

Due to its construction, the PCF is also defined as multimode step index fiber.

From the spectral characteristic of the attenuation coefficient (see Fig. 12.2), an attenuation of 8–10 dB/km can be read in the range from 650–660 nm, which is of interest for plastic optical fibers. The optical attenuation of this fiber therefore is several times less than that of the plastic optical fiber.

By contrast there is a significantly smaller fiber core cross-section of 200 μm compared with the plastic optical fiber with 980 μm. As a result, a higher coupling loss is to be expected from transmitters and receivers that are adapted to the conditions of plastic optical fibers. By means of a simple calculation, as shown below, this expectation is confirmed.

A transmitter for plastic optical fibers typically has an emitting area of about 0.3 mm · 0.3 mm = 0.09 mm². The amount of power that can be launched for such a transmitter is known. The 200 μm PCF, however,

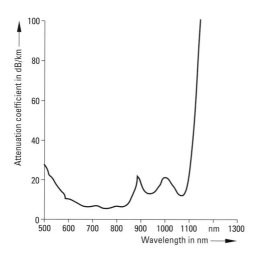

**Fig. 12.2** Spectral characteristic of the attenuation coefficient $\alpha(\lambda)$ of a PCF

only has a cross-sectional area of 0.031 mm². This means that no more than 34.5% of the emitted light power can be launched into the PCF. In comparison with the power for a plastic optical fiber, therefore, 4.6 dB less light is launched into the PCF. Taking into account the smaller numerical aperture of the PCF (0.36), the light power launched into the optical fiber is about 1 dB less compared to the plastic optical fiber. Higher coupling losses in the order of 9 dB are often encountered in practice. The reasons for this are associated with the coupling conditions such as separation and angle of the PCF to the LED etc. These properties are dependent on the special transmitter and connector system and must be examined in practice. Nevertheless, as the following example shows, distances of about 500 m can usually be spanned using PCF.

Example:

A system designed for plastic optical fiber permits a bridgeable distance of $L_{max} = 60$ m using plastic optical fiber (at 660 nm). This assumes a fiber loss of $\alpha_{POF} = 230$ dB/km for the POF. What distance can be spanned if the coupling loss is 9 dB and the attenuation coefficient is 10 dB/km?

Using the specifications for the plastic optical fiber, first calculate the available power budget $P_{budget}$:

$$P_{Budget} = L_{max} \cdot \alpha_{POF}$$

$$P_{Budget} = 60 \text{ m} \cdot 230 \text{ dB/km}$$

$$P_{Budget} = 13.8 \text{ dB}$$

This power budget is now also available for the PCF, for which the coupling loss of 9 dB must be taken into account.

$$P_{Budget} - 9 \text{ dB} = L_{max} \cdot \alpha_{PCF}$$

$$\frac{13.8 \text{ dB} - 9 \text{ dB}}{10 \text{ dB/km}} = L_{max}$$

$$L_{max} = 0.48 \text{ km}$$

This shows that distances of up to 480 m can be spanned with this system. A sufficient system reserve should also be considered in such cases (see section 13.1)

## 12.3 Cables using PCF

The simplest construction is again a single buffered fibre, shown in Fig. 12.3 as a simplex and duplex buffered fibre. In contrast to the POF buffered fibre, however, the PCF core already has strength members. Buffers

Simplex           Duplex

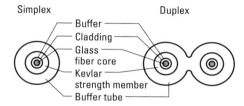

**Fig. 12.3**    PCF simplex and duplex core

with external diameters of 2.2 mm and 3.0 mm are frequently used. The same materials are used for the PCF buffer as for POF, and again there is a trend toward halogen-free and flame-resistant materials.

Buffered PCF are available with the following typical characteristics:

| | |
|---|---|
| Core diameter: | 200 μm |
| Cladding diameter: | 230 μm |
| Buffer diameter: | 500 μm |
| Tensile strength: | 45 N |
| Minimum bending radius: | 16 mm |
| Attenuation coefficient at 660 nm: | 10 dB/km |
| Numerical aperture: | 0.36 |
| Acceptance angle: | 21° |
| Bandwidth-distance product: | 17 MHz · 1 km |

## 12.4 Cable assembly and connector systems

The connectorizing of PCFs is made very simple by the use of a special "crimp and cleave" technology and can be carried out in about three minutes. This makes the fibers particularly suitable for on-site assembly. In this procedure the connector is simply crimped onto the fiber without the need for adhesive. The end face of the fiber is processed by a special cleaving tool, which scores the fiber under tension and then cleaves (breaks) it. Grinding and polishing is not necessary as with conventional glass fibers.

All the connector systems available for plastic optical fibers (see section 9.3) can also be considered for use with PCF. In addition, other familiar connectors can in many cases be adapted to these special fibers. The use of crimpless connectors, in which the connector is fitted to the cable without the need for additional tools, is only a question of time as development work is already underway.

# 13 System development

Any user who want to construct a transmission link by using plastic optical fibers must know the following parameters:

▷ Minimum and maximum distance to be spanned

▷ Data rate

▷ Operating wavelengths

▷ Minimum and maximum operating temperature of a system

▷ Type and properties of the cable to be used

▷ Any required number of passive coupling points, if necessary

This data is the basis for planning the installation from a transmission viewpoint. In addition, there are other questions to be considered such as cost, connector type etc. when planning a system, but which will not be considered in further detail here.

## 13.1 Attenuation plan

The total loss of a system $A_{Sys}$ comprises the cable length loss, the coupling loss at transmitter $A_{CT}$ and receiver $A_{CR}$ and the coupling loss at any passive couplings $A_{CP}$ together with any loss margin which must be allowed for $A_{Mar}$. This produces the following equation:

$$A_{Sys} = \alpha_{OF} \cdot L + A_{CT} + A_{CR} + n \cdot A_{CP} + A_{Mar} \qquad (13.1)$$

$A_{Sys}$ Total system loss (attenuation) in dB
$\alpha_{OF}$ Attenuation coefficient of the cable in the dB/km
$A_{CT}$ Input coupling loss at transmitter
$A_{CR}$ Output coupling loss at receiver
$A_{CP}$ Coupling loss at passive coupling
$A_{Mar}$ Loss margin (e.g. 3 dB)
$L$ Lengths of cable in km
$n$ Number of passive couplings

With most commercially available transmitters and receivers the coupling loss does not have to be considered, as the manufacturers test the launch power or receiver sensitivity with the aid of connectorized plastic optical fibers and specify them in the data sheet. In this way, the losses are already accounted for in advance.

126

As described in Chapter 4, the attenuation coefficient of the plastic optical fiber is a length-dependent parameter when using LEDs as transmitters. Usually, however, only the attenuation coefficient, measured monochromatically, is known. Occasionally the manufacturer's specifications contain guide values for the attenuation, measured with an LED, based on a wavelength of 660 nm and a fiber length of 50 m, which can be used for planning purposes. If no information is available, it is recommended that the attenuation coefficients are determined using the special transmitter. If a system is designed for a length of 50 m, then the specification of the attenuation coefficient relative to 50 m is totally sufficient. Although, due to the non-linearity of the attenuation coefficients (see Fig. 4.4), the attenuation coefficient for shorter distances below 50 m is greater than that for a length of 50 m, the absolute loss of the 50 m cable is the largest of all the losses that occur in each case. When planning systems that extend beyond 50 m, calculations can still be made with attenuation coefficients based on 50 m. In no circumstances is any risk taken. If, however, the full benefit is to be gained from the system, more accurate investigations must be performed on the cable.

Example:

The system loss must be determined for a system whose maximum dimension is 50 m. A passive adapter must be provided. The mean loss of the adapter is $A_{CP} = 3$ dB. In addition, a system margin of 3 dB must be allowed for. The operating wavelength is $\lambda = 660$ nm. With this wavelength and a cable length of 50 m, the attenuation coefficient is 230 dB/km. From these specifications, the overall system loss $A_{sys}$ is obtained according to equation 13.1:

$$A_{Sys} = 230 \text{ dB/km} \cdot 0.050 \text{ km} + 1 \cdot 3 \text{ dB} + 3 \text{ dB} = 17.5 \text{ dB}$$

In other words, a power budget of at least 17.5 dB is necessary in order to implement this system.

This power budget must be achieved by the appropriate selection of transmitters and receivers. The following considerations refer in each case to receivers that already emit a TTL signal at the output. When using receivers that only supply analog signals, it is left to the user to set up an appropriate transducer connection. The data for this connection can be used for the observations below.

To calculate the power budget, the values from the manufacturer's data sheets can be used. The values are usually specified in dBm, making the calculation very simple. Typically the minimum ($P_{Tmin}$), the typical ($P_{Ttyp}$) and the maximum ($P_{Tmax}$) optical transmitter powers are specified for the transmitter. In the worst case scenario, $P_{Tmin}$ should be used for calculating the power budget. The receiver sensitivity is specified in the same way. Whereas these specifications refer to the scattering of the op-

127

tical transmitter power for the transmitter, they denote the dynamic range in the case of the receiver. The dynamic range refers to the difference between the minimum receiver sensitivity and maximum receiver sensitivity. The minimum receiver sensitivity refers to the exact minimum light power that the receiver can still detect and process. The maximum receiver sensitivity is the maximum light power that the receiver can accept without overdriving. For this reason, the maximum receiver sensitivity is often referred to as the overdrive limit. The calculation of the power budget is performed as follows:

$$P_{\text{Budget}} = P_{\text{Tmin}} - P_{\text{Rmin}} \tag{13.2}$$

$P_{\text{Tmin}}$  minimum transmit power in dBm
$P_{\text{Rmin}}$  minimum receiver sensitivity in dBm

From time to time the minimum and maximum receiver sensitivity is specified in datasheets with reference to the TTL level of the electrical output signal. In this case, $P_{\text{Rmin}}$ should be replaced by $P_{\text{Rmin(Low)}}$ in formula 13.2, where $P_{\text{Rmin(Low)}}$ means the minimum input power that is necessary to obtain a LOW signal at the output. These relationships are shown again in Fig. 13.1 for clearer understanding.

If $P_{\text{Tmin}} > P_{\text{Rmax}}$, then there is a risk that the receiver will be overdriven when using a short cable. In this case, either a longer cable is used (this

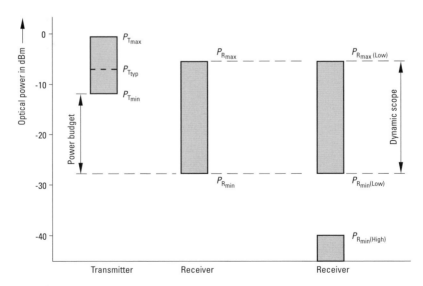

**Fig. 13.1**
Graphical explanation of the terms "power budget" and "dynamic range"

is quite common in practice), or the optical output power must be reduced via the current at the transmitter, so that over driving is avoided. The simplest solution however, is to rule this possibility out by selecting the appropriate transmitter and receiver.

In order that the overall system functions when the longest cable is being used, the following condition must be met:

$$A_{Sys} \leq P_{Budget} \tag{13.3}$$

Example:

A transmitter and a receiver with the following optical data are available. The calculation should be made to determine the maximum distance that can be bridged with a plastic optical fiber, taking into account the system margin of 3 dB.

Optical data for the transmitter and receiver:

$P_{Tmax}$  =  $-4$ dBm
$P_{Tmin}$  =  $-6$ dBm
$P_{Rmax}$  =  $-6$ dBm
$P_{Rmin}$  =  $-28$ dBm
$\lambda$  = 660 nm

First we check whether $P_{Tmin}$ is smaller or equal to $P_{Rmax}$. As these values are the same, the system will not be overdriven, even with short cables. From the data sheet of the plastic optical fiber we determine that the attenuation coefficient is 230 dB/km when using an LED with a wavelength of 660 nm, based on a length of 50 m. The maximum bridgeable distance is calculated as follows:

$$P_{Budget} = -6 \text{ dBm} - (-28 \text{ dBm}) = 22 \text{ dB}$$

$$A_{Sys} = 230 \text{ dB/km} \cdot L_{max} + 3 \text{ dB} = P_{Budget}$$

$$L_{max} = \frac{22 \text{ dB} - 3 \text{ dB}}{230 \text{ dB/km}} \approx 82 \text{ m}$$

In this case therefore we can span a distance of up to 82 m using POF.

As shown in sections 8.2.1 and 8.3, the wavelength, output power of the transmitter and receiver sensitivity all depend on the temperature. If these dependencies have not already been included in the specified optical data of the transmitter and receiver, this must be carried out afterwards.

Example:

Transmitters and receivers from the above example should be used in the temperature range from 25 °C to 70 °C. The following data is available for $T = 25$ °C:

$\Delta\lambda/\Delta T = 0.12 \ nm/K$
$\Delta P_T/\Delta T = -0.04 \ dB/K$
$\Delta P_R/\Delta T = +0.03 \ dB/K$

What is the maximum distance that can be bridged taking into account the temperature?

a) Shifting the transmitter wavelength:

$$\lambda_{70\,°C} = \lambda_{25\,°C} + \frac{\Delta\lambda}{\Delta T} \cdot (T_{max} - T)$$

$$\lambda_{70\,°C} = 660 \ nm + 0.12 \ nm/K \cdot 45 \ K = 665.4 \ nm$$

As a guide value for the attenuation coefficient of the POF at this wavelength, we use the specifications from Table 4.3. According to these, an attenuation increase of about 20 dB/km is to be expected when shifting the wavelength from 660 nm to 665.4 nm. This increases the attenuation coefficient from an initial 230 dB/km to 250 dB/km in our example.

b) Changing the optical output power and the receiver sensitivity:

$$P_{Rmin\ 70\,°C} = P_{Rmin\ 25\,°C} + \frac{\Delta P_R}{\Delta T} \cdot (T_{max} - T)$$

$$P_{Rmin\ 70\,°C} = -28 \ dBm + 0.03 \ dB/K \cdot 45 \ K$$

$$P_{Rmin\ 70\,°C} = -26.65 \ dBm$$

$$P_{Tmin\ 70\,°C} = P_{Tmin\ 25\,°C} + \frac{\Delta P_T}{\Delta T} \cdot (T_{max} - T)$$

$$P_{Tmin\ 70\,°C} = -6 \ dBm + (-0.04 \ dB/K) \cdot 45 \ K$$

$$P_{Tmin\ 70\,°C} = -7.8 \ dBm$$

c) Calculation of the maximum bridgeable distance:

$$L_{max} = \frac{P_{Tmin\ 70\,°C} - P_{Rmin\ 70\,°C} - A_{Res}}{\alpha_{OF\ 665.4\ nm}}$$

$$L_{max} = \frac{-7.8 \ dBm + 26.65 \ dBm - 3 \ dB}{250 \ dB/km}$$

$$L_{max} \approx 63 \ m$$

In the temperature range from 25 °C to 70 °C therefore, a maximum distance of 63 m can be bridged under the conditions specified. The smallest bridgeable distance is to be determined in the same way.

In conclusion, it should be noted that the peak wavelengths of the LED exhibit scattering as a result of the manufacturing process. When performing the calculation, the worst case should be assumed, i.e. assume the wavelengths at which the plastic optical fiber shows the greatest loss.

## 13.2 Data rate

The maximum transmittable data rate depends firstly on the active components, i.e. the transmitter and receiver, and secondly on the bandwidth of the plastic optical fiber. This data can be obtained from the manufacturer's specifications. According to Formula 1.17, the product of bandwidth and length of the optical fiber is approximately constant. We therefore assume that the bandwidth decreases on a linear basis with the length of the optical fiber, which is sufficiently accurate for the relatively short lengths of POF. Accordingly, the following conditions must be met when planning a system:

$$B_{sys} \geq \frac{B \cdot L}{L_{max}} \tag{13.4}$$

$B_{sys}$   System bandwidth
$B \cdot L$   Bandwidth-distance product of the plastic optical fiber
$L_{max}$   Maximum bridgeable distance

Example:

A system is to be constructed with a data rate of 150 Mbit/s. What is the maximum distance that can be bridged with a plastic optical fiber whose numerical aperture $NA = 0.47$? The optical transmitter launches light with an $NA$ of 0.65.

As a rule of thumb, it may be assumed that for each bit per second 0.8 Hz is required as transmission speed (applies to NRZ coding).

From Table 4.4 we assume that the plastic optical fiber under these conditions has a bandwidth-distance product of approximately 45 MHz · 100 m. The maximum bridgeable distance is thus calculated as follows:

$$0.8 \cdot 150 \text{ MHz} = \frac{45 \text{ MHz} \cdot 100 \text{ m}}{L_{max}}$$

$$L_{max} = 37.5 \text{ m}$$

This means that a maximum distance of 37.5 m can be bridged. In this case the use of low $NA$ fiber is recommended. The bandwidth-distance product of this fiber under the conditions specified is approximately 110 MHz · 100 m. This would give a maximum bridgeable distance of 91 m.

# 14 Standardization

The development of standards in the field of optical fiber technology and in particular for plastic optical fiber technology is being stepped up around the world. Whereas the standardization of glass optical fibers is at an advanced stage, the standardization for plastic optical fibers only began between six and eight years ago.

## 14.1 Standardization committees

There is a series of organizations involved in the development of standards for plastic optical fibers:

▷ International Electrotechnical Commission (IEC): responsible for the international standardization in the field of electrotechnology and electronics.

▷ American National standards Institute (ANSI): coordinates the national standardization activities and represents the USA in questions of international standardization.

▷ Institute of Electrical and Electronic Engineers (IEEE): responsibilities include development of LAN standards for data processing modules.

▷ *Deutsches Institut für Normung e.V.* (DIN) = German Institute of Standardization

▷ *Verband Deutscher Elektrotechniker e.V* (VDE) = Association of German Electrical Engineers

The authoritative international standardization institution is the IEC, based in Geneva. The major committees of the IEC concerned with the subject of "fiber optics" are as follows:

▷ TC86        Fiber optics

▷ SE86A       Fibers and cables

▷ SWG1        Short links: fiber cable

▷ SWG2        Applications

▷ SC86B       Connecting and passive components

▷ WG1         Connections

▷ WG2         Passive components

▷ SC86C      System specification

▷ TC83/WG2   Local networks

▷ TC47/WG15  Optical components

The main activities on standardization of plastic optical fibers are carried out in Working Group 802 of the IEEE.

As most activities on development and manufacture of plastic optical fiber originated in Japan, standards for plastic optical fibers are also found in the Japanese Industry Standard (JIS).

## 14.2 Standards for fiber, cable and connectors

JIS and IEC specify the test procedure for the mechanical characterization, the structural parameters and the attenuation of the multimode POFs. The standards refer to fiber diameters of 0.5 mm, 0.75 mm and 1 mm. In general, only minor differences exist between the corresponding JIS or IEC standards. The problem areas and the corresponding standards are compared in Table 14.1

The test procedures for the characterization of fiber cores and cables are explained in Chapter 7 of this book. The product specifications for plastic optical fibers are listed in Table 14.2 and those for plastic optical fiber cores in Table 14.3

Standardized connectors for plastic optical fibers are described in Section 9.3 of this book (see Table 9.1).

## 14.3 Standards for POF systems

*SERCOS* interface is an open standard interface for communication between controllers and digital drives in numerically controlled machines. *SERCOS interface* is standardized as IEC 601491 SYSTEM international standard and defines plastic optical fiber as the only transmission medium.

**Table 14.1**
Test procedures and products specifications conforming to JIS and IEC

| Subject | JIS | IEC |
|---|---|---|
| Test procedures, fibers | JIS C 6863<br>JIS C 6862 | 60793-1<br>– |
| Product specification, fibers | JIS C 6837 | 60793-2 |
| Test procedures, cables | – | 60794-1 |
| Product specification, cables | JIS C 6836 | – |

**Table 14.2**   Product specification for plastic optical fibers

| Subject | JIS | | | IEC |
|---|---|---|---|---|
| Refractive index profile | Step index | | | as JIS |
| Core diameter in μm | 980 | 735 | 485[1] | as JIS |
| Cladding diameter in μm | 1000 ± 60 | 750 ± 45 | 500 ± 30 | as JIS |
| Deviation from circular form in cladding surface as percentage | ≤6 | ≤6 | ≤6 | as JIS |
| Max. theoretical numerical aperture | 0.5 ± 0.15 | | | as JIS |
| Attenuation | ≤0.3 dB/m [2] [3] ≤0.4 dB/m [2] | | | ≤0.3 dB/m [3] ≤0.4 dB/m |
| Bandwidth | – | – | – | ≥10 MHz · 100 m |

[1] Nominal values: normally the core diameter is 10 to 20 μm less than the cladding diameter
[2] The wavelength of the source should be (650 ± 3) nm (JIS C 6863)
[3] When launching with mode equilibrium.

**Table 14.3**   Product specification for plastic optical fiber cores

| Subject | JIS | | | IEC |
|---|---|---|---|---|
| Refractive index profile | Step index | | | as JIS |
| Core diameter in μm | 980 | 735 | 485[1] | as JIS |
| Cladding diameter in μm | 1000 ± 60 | 750 ± 45 | 500 ± 30 | as JIS |
| Buffer diameter in mm | 2.2 ± 0,1 | 2.2 ± 0,1 | 1.5 ± 0.1 | – |
| Deviation from circular form in cladding surface as % | ≤6 | ≤6 | ≤6 | as JIS |
| Max. theoretical numerical aperture | – | – | – | 0.5 ± 0.15 |
| Numerical aperture | 0.5 ± 0.15 | | | – |
| Attenuation | as in Table 14.2 | | | as JIS |
| Bandwidth | – | | | ≥10 MHz · 100 m |

[1] Nominal values

*INTERBUS-S* is a sensor/actuator bus and specifies both POF and PCF as transmission media in DIN EN 19258.

*PROFIBUS* is a field bus that is standardized in EN 50170, Volume 2. Both plastic optical fibers and glass optical fibers are specified as transmission media.

Apart from these existing system specifications, further activities are in progress which are aimed at specifying plastic optical fiber as a transmission medium. In May 1997, for example, the ATM Forum approved a specification for the use of plastic optical fiber for distances up to 50 m at a data rate of 155 Mbit/s. PCF is specified as a transmission medium for distances up to 100 m.

The use of plastic optical fiber as a potential transmission medium has also been discussed during the specification of IEEE 1394 (high-speed serial bus).

## 14.4 National workgroups

National workgroups are currently carrying out work in Japan, the USA, France and Germany, whose aim is to develop, distribute and support standardization procedures for POF products.

In Germany these activities are performed in Technical Group 5.4.1 of the ITG (= Information Technology Association) which is part of the VDE.

In Japan a consortium of 61 companies and institutes is currently working in the field of plastic optical fibers.

The USA is involved in a special interest group for plastic optical fibers, while in France between 50 and 70 scientists meet twice a year under the leadership of the French Plastic Optical Fiber Club (CFOP), in order to exchange information about the latest developments.

Between 100 and 200 scientists and engineers meet once a year at the international POF Conference to discuss the results of new developments.

# 15 Appendix

## 15.1 Format of type codes for plastic optical fibers consistent with DIN VDE 0888

$\square - \square\square\square\square\square \quad \square\square\square/\square \quad \square\square\square \quad \square$

1 − 2 3 4 5 6   7 8 9   10   11 12 13   14

| 1 | I | Indoor cable |
|---|---|---|
| | A | Outdoor cable |
| | AT | Outdoor fan-out cable |
| 2 | F | Fiber |
| | V | Tight buffered fiber |
| | H | Single-fiber loose buffer |
| | W | Single-fiber loose buffer, filled |
| | B | Multifiber loose buffer |
| | D | Multifiber loose buffer, filled |
| 3 | S | Metallic element in the cable core |
| 4 | F | Filling compound of cable core stranding cavities |
| 5 | H | Sheath of halogen-free material |
| | Y | Polyvinyl chloride (PVC) sheath |
| | 2 Y | Polyethylene (PE) sheath |
| | 11 Y | Polyurethane (PUR) sheath |
| | 4 Y | Polyamide (PA) sheath |
| | 2 X | VPE sheath |
| | (L) 2 Y | Laminated sheath |
| | (D) 2 Y | Polyethylene sheath with plastic barrier foil |
| | (ZN) 2 Y | Polyethylene sheath with non-metallic strength member |
| | (L) (ZN) 2 Y | Laminated sheath with non-metallic strength member |
| | (D) (ZN) 2 Y | Polyethylene sheath with plastic barrier foil and non-metallic strength member |
| 6 | H | Inner sheath of halogen-free material |
| | Y | Inner sheath of PVC |
| | 2 Y | Inner sheath of PE |
| | 11 Y | Inner sheath of PUR |
| | 4 Y | Inner sheath of PA |
| | B | Armoring |
| | BY | Armoring with PVC protective cover (e.g. rodent protection) |
| | B2 Y | Armoring with PE protective cover (e.g. rodent protection) |

| | | |
|---|---|---|
| | B 11 Y | Armoring with PUR protective cover (e.g. rodent protection) |
| 7 | ..x.. | Number of buffered fibers or multifiber units × number of fibers per unit |
| 8 | E | Single-mode fiber |
| | G | Graded index fiber |
| | S | Step index fiber (glass/glass) |
| | K | Step index fiber (glass/plastic) |
| | Q | Quasi-graded index fiber |
| | P | Polymer fiber |
| 9 | .. | Core diameter in μm |
| 10 | .. | Cladding diameter in μm |
| 11 | .. | Attenuation coefficient in dB/km |
| 12 | .. | Wavelength |
| | | A = 650 nm |
| | | B = 850 nm |
| | | F = 1300 nm |
| | | H = 1550 nm |
| 13 | .. | bandwidth-distance product in MHz · km or MHz · 100 m for plastic optical fiber |
| 14 | Lg | Stranding in layers |

Note:
According to DIN VDE, position 6 – Inner sheath is only provided for outdoor cable. This position is increasingly being specified for indoor cables in order to represent the protective sheath material of the optical fiber elements. This makes the designation of indoor cables clearer. This method of representation has been used in this case.

Hybrid cables

For positions 1–14, see above; these are followed without a space by position 15–17:

| | | |
|---|---|---|
| 15 | + .. × .. | + number of buffered fibers × conductor cross section |
| 16 | F-Cu | Fine-wire copper conductor |
| | FF-Cu | Very fine wire copper conductor |
| 17 | U/$U_o$V | Rated voltage |

## 15.2 Symbols and abbreviations used in equations

| | |
|---|---|
| $\alpha$ | Attenuation coefficient in dB/km; |
| $\alpha$ | Stranding angle |
| $\Delta$ | Normalized refractive index difference |
| $\Delta\tau$ | Delay difference |
| $\lambda$ | Wavelength |

137

| | |
|---|---|
| $\lambda_{peak}$ | peak wavelength |
| $\theta$ | Angle relative to axis of optical fiber |
| $\theta_{max}$ | Acceptance angle |
| $\Omega$ | Space angle |
| | |
| A | Cross-sectional area |
| A | Attenuation in dB |
| $a$ | Core radius |
| $A_{CP}$ | Coupling loss at passive coupling |
| $A_{CR}$ | Output coupling loss at receiver |
| $A_{CT}$ | Input coupling loss at transmitter |
| $A_{Mar}$ | Loss margin |
| $A_{Sys}$ | Total system loss in dB |
| B | Bandwidth |
| $B_{sys}$ | System bandwidth |
| c | Speed of light in a vacuum |
| $CR$ | Coupling ratio |
| dB | Decibel |
| $E_g$ | Energy gap |
| $E_L$ | Energy conduction band |
| $EL$ | Excess loss |
| eV | Electron volt |
| $E_v$ | Energy valence band |
| $f$ | Frequency |
| FWHM | Full width at half maximum |
| g | Profile exponent |
| $g_{opt}$ | Optimized profile exponent |
| h | Planckian efficiency quantum |
| $I$ | Current |
| $IL$ | Insertion loss |
| $I_s$ | Threshold current |
| km | Kilometer |
| $L$ | Length of transmission link |
| $L$ | Length of stranding element |
| $L$ | Radiance |
| LAN | Local area network |
| LD | Laser diode |
| LED | Light emitting diode |
| $L_{max}$ | Maximum bridgeable distance |
| MQW | Multiple quantum well |
| $n$ | Refractive index |
| NA | Numerical aperture |
| $n_c$ | Core refractive index |
| $n_{gr}$ | Group refractive index |
| NRZ | Non return to zero |
| OF | Optical fiber |

| $P$ | Power |
|---|---|
| $P_{opt}$ | Optical power |
| $P_{Rmin}$ | Minimum receiver sensitivity |
| $P_{Tmin}$ | Minimum transmit power in dBm |
| POF | Plastic optical fiber |
| PCF | Polymer cladded fiber |
| PMMA | Polymethylmethacrylate |
| $R$ | Coefficient of reflection |
| $R$ | Stranding radius |
| $r$ | Radius of curvature |
| $s$ | Distance |
| $S$ | Stranding pitch |
| $T$ | Transmission coefficient |
| U | Uniformity |
| $w_o$ | Mode field radius in single-mode optical fiber |
| $Z$ | Excess length due to stranding |

# 16 Bibliography

[1.1] Ritter, M. B.: Dispersion Limits in Large Core Fibres. Proc. POF '93, International Conference. European Institut for Communications and Networks, AKM Messen AG, 1993, P. 31–34

[1.2] D. Gloge et al: Multimode theory of graded core fibers. Bell System Tech. Journal 52 (1973). No. 9, P. 1563–1578.

[3.1] Kämpf, G.; Freitag, D.; Witt, W.: Polycarbonat und Licht, Angewandte Makromolekulare Chemie 183 (1990), P. 243–272

[3.2] Kaino; T.: Ultimate loss limit estimation of plastic optical fibers, Kobunshi Robunshu, 42 (1985). No. 4, P. 257–264

[3.3] Koike, Y.: Status of POF in Japan. Proc. POF '96, International Conference. Institut of Communications, AKM Messen AG, 1996, P. 1–7

[3.4] Koike, Y.: Design of core diameter of wide-band GI-POF. Proc. POF '95, International Conference. European Institute of Communications and Networks, 1995, P. 92–99

[3.5] Koike, Y.; Ishigure, T.; Nihei, E.: Journal of Ligthwave Technology 13 (1995). No.7

[3.6] Ishigure, T.; Nihei, E.; Yamazaki, S.; Kobayashi, K.; Koike, Y.: Electron. Lett. 31 (1995) P. 467

[3.7] Ziemann, O.: Grundlagen und Anwendungen optischer Fasern. Der Fernmeldeingenieur 11/12 (1996)

[3.8] Koike, Y.; Nihei, E.: Single Mode Polymer optical fiber. Design Manual and Handbook & Buyers Guide. Boston: Information Gatekeepers, Inc., 1993

[3.9] Ashpole, R. S.; Hall, S. R.; Luker, P. A.: Polymer optical fibers, A case study. Boston: Information Gatekeepers, Inc., 1993

[4.1] Takahashi, S.: Experimental studies on launching conditions in evaluating transmission characteristics of POF's. Proc. POF '93, International Conference. European Institut for Communications and Networks, AKM Messen AG, 1993, P. 83–85

[4.2] Yoshimura, T.; Nakamura, K.; Okita, A.; Nyu, T.; Yamazak, S.; Dutta, A. K.: Experiments on 155 Mbps 100 m Transmission using

650 nm LED and Step Index POF. Proc. POF '95, International Conference. European Institute of Communications and Networks, 1995, P. 119−121.

[5.1] Goehlich, L.; Scharschmidt, J.: Offenen Kommunikation. EV-Report 1 (1992), P. 25−28

[8.1] Bludau, W.: Halbleiter-Optoelektronik. München: Carl Hanser 1995.

[8.2] H. Sugawara et al: High-Efficiency InGaAlP Green Light-Emitting Diodes. Abstracts of the 1991 International Conference on Solid State Devices and Materials, Yokohama 1991.

[8.3] Fukuoka, K.; Iwakami, T.; Schumacher, K.: High-speed and long-distance POF transmission system based on LED transmitter. Proc. POF '93, International Conference. European Institut for Communications and Networks, AKM Messen AG, 1993, P. 43−45.

[8.4] Krumpholz, O.; Pressmar, K.; Schlosser, E.: LED-carrier with reflector for plastic optical fibers. Proc. POF'93, International Conference. European Institut for Communications and Networks, AKM Messen AG, 1993, P. 125−126

[8.5] Yoshimura, T.; Nakamura, K.; Okita, A.; Nyu, T.; Yamazaki S.; Dutta, K.: 1995. Experiments on 156 Mbps 100 m Transmission using 650 nm LED and Step Index POF. Proc. POF '95, International Conference. European Institut for Communications and Networks, 1995, P. 119−121

[8.6] S. Yamazaki et al: High Speed Plastic Fiber Transmission for Data Communication. International Conference. Symposium on New Trend or Advanced Materials. The Society of Polymer Science, 1995, P. 1−4.

[8.7] D. M. Kuchta et al: High Speed Data Communication Using 670 nm Vertical Cavity Surface Emitting Lasers and Plastic Optical Fibers. Proc. POF '94, International Conference. European Institut for Communications and Networks 1994, P. 135−139

[8.8] F. Miyaska et al: High-speed light souces and detectors for gigabit-per-second plastic-optical-fiber transmission. Papers on high Data Rate Applications Using Graded-Index (GI) and Step-Index (SI) Plastic Optical Fibers (POF). Information Gatekeepers, Inc. 1996, P. 65−66.

[8.9] Koike, Y: High-Speed Multimedia POF Network. Proc. POF '94, International Conference. European Institut for Communications and Networks, P. 16−20.

[10.1] Schreiter, G.; Hotea, G.; Engel, A.: Fibre optic connectors for consumer applications. Proc. POF '93, International Conference. European Institut for Communications and Networks, AKM Messen AG, 1993, P. 132–135.

[10.2] Cirillo, J.R.; Jennings, K.L.; Lynn, M.A.: Connection system designed for plastic optical fiber local area networks. Plastic Optical Fibers, Mototaka Kitatawa, John F. Kreidl, Robert E. Steel, SPIE 1592, 1991, P. 53–59.

[11.1] Kalymnios, D.: Plastic optical fibre tree couplers using simple Y-couplers. Proc. POF '92, International Conference. IGI Europe, AKM Messen AG, 1992, P. 115–118.

[11.2] Yuuki,H.; Ito, T.; Sugimoto, T.: Plastic star coupler. Plastic Optical Fibers, Mototaka Kitatawa, John F. Kreidl, Robert E. Steel, SPIE 1592, 1991, P. 2–11.

[11.3] Marcou, J.; Faugeras, P.: Interconnection components for plastic optical fibers. Proc. POF '92, International Conference. IGI Europe, AKM Messen AG, 1992, P. 109–114.

[11.4] E. Th. C. van Woesik et al: N×N Bi-directional transmissive star coupler. Proc. POF '93, International Conference. European Institut for Communications and Networks, AKM Messen AG, 1993, P. 127–131.

[11.5] Eickhoff, W.; Haag, H.G.; Stankovic, D.; Zamzow, P.E.: Polymer optical fiber cables for both automotive and customer premises application. Plastic Optical Fibers and Applications. Information Gatekeepers, Inc., 25, 1990, P. 162–169.

[11.6] Rogner,A.: Micromoulding of passive network components. Proc. POF '92, International Conference. IGI Europe, AKM Messen AG, 1992, P. 102–104.

[11.7] Rogner, A.; Pannhoff, H.: Characterization and qualification of moulded couplers for POF-networks. Proc. POF '93, International Conference. European Institut for Communications and Networks, AKM Messen AG, 1993, P. 136–139.

[11.8] Saitoh, N.; Shimada, K.: Plastic optical fibres and their application to passive components and various digital data links. Proc. POF'92, International Conference. IGI Europe, AKM Messen AG, 1992, P. 10–14.

# Index

Mahlke, Günter / Gössing, Peter
# Fiber Optic Cables
**Fundamentals Cable Engineering System Planning**

3rd extensively revised and enlarged edition, 1997, 276 pages, 175
illustrations, 37 tables, hardcover
ISBN 3-89578-068-5
DM 119,00 / € 60,84 / sFr 105,00

This publication is directed towards all who deal with design, construction and
maintenance of fiber optic cable plants. Furthermore, it provides basic information as an introduction to specialized technical literature.

In order to make it easier to study the many specialized publications, the book
contains a detailed glossary of technical terms.

For the new edition, the major developments over the last years have been taken
into account. As transmission rates are rapidly growing, the part on the fundamental principles has been updated with a section on nonlinear optical effects.
New chapters cover tight buffer cables for customer premise cabling as well as
submarine cables for non-repeatered applications and aerial cables for use by
power companies. In addition, the section concerning accessory equipment like
splices, connectors and closures has been considerably enlarged.

## Contents

Historical Development · Physics of Optical Waveguides · Chemistry of Optical
Waveguides · Optical Waveguide Profiles · Optical Fiber Parameters and
Measurement Methods · Optical Fiber Construction · Manufacturing of Optical
Fibers · Optical Fiber Buffers · Optical Cable Design · Cable Plant Design ·
Network Configurations · Electrooptic Signal Conversion · Fiber Optic
Components · Systems for Optical Transmission.

Heublein, Hans
# Transmitting Data without Interference
**Cables in Building Installations and Industrial Measurement and Process
Control**

1998, 118 pages, 36 illustrations, 18 tables, hardcover
ISBN 3-89578-073-1
DM 59,00 / € 30,17 / sFr 53,00

Power supply applications found within building installations and in industrial
measurement and process control increasingly overlap with those of communications technology. Therefore, planners and electricians must understand and consider the particular fundamentals of both disciplines.

This book comprehensively explains the various mechanisms of interference arising during signal and data transmission between sender and receiver modules as
well as within the entire transmission system. It concentrates, thereby, largely on
the aspects of electromagnetic compatibility.

Berger, Hans
# Automating with SIMATIC S7
**Integrated Automation with SIMATIC S7-300/400**
Controllers, Software, Programming, Data Communication, Operator
Control and Process Monitoring

2000, approx. 250 pages, 100 illustrations, 20 tables, hardcover
ISBN 3-89578-133-9
DM 88,00 / € 44,99 / sFr 80,00

For the example of the SIMATIC S7 programmable controller, the reader is given
an overview of the functioning and design of a modern automation system, an
insight into the configuring and parameterization of hardware with STEP 7 and
the solution of control problems with different PLC programming languages.


Bezner, Hans
# Dictionary of Electrical Engineering, Power Engineering and Automation
**(Part 1)**
4th revised and enlarged edition, 1998, 608 pages, hardcover
ISBN 3-89578-077-4
DM 128,00 DM / € 65,45 / sFr 113,00

**(Part 2)**
4th revised and enlarged edition, 1998, 532 pages, hardcover
ISBN 3-89578-079-0
DM 128,00 DM / € 65,45 / sFr 113,00

The previous "Dictionary of Power Engineering and Automation" is a standard
work for all those requiring a comprehensive and reliable compilation of terms
from the fields of power generation, transmission and distribution, drive engineering, switchgear and installation engineering, power electronics, measurement, analysis and test engineering as well as automation engineering. Together
with the great number of basic electrotechnical terms, it comprehensively covers
large fields of industrially applied electrical engineering with its more than 67,000
entries in Volume 1 (German/English) and 54,000 entries in Volume 2 (English/
German). This universal character is also clearly shown by the new title.

Berger, Hans
# Automating with STEP 7 in STL and SCL
## SIMATIC S7-300/400 Programmable Controllers

2000, approx. 430 pages, 180 illustrations, 120 tables, hardcover
ISBN 3-89578-140-1
DM 128,00 / € 65,45 / sFr 113,00

This book describes elements and applications of the graphic-oriented programming languages LAD (ladder diagram) and FBD (function block diagram) for use with both SIMATIC S7-300 and SIMATIC S7-400. It is aimed at all users of SIMATIC S7 programmable controllers. First-time users will be introduced to the field of programmable logic control whereas advanced users will learn about specific applications of SIMATIC S7 programmable controllers.


Berger, Hans
# Automating with STEP 7 in LAD and FBD
## SIMATIC S7-300/400 Programmable Controllers

2000, approx. 370 pages, 150 illustrations, 100 tables, hardcover
ISBN 3-89578-131-2
DM 128,00 / € 65,45 / sFr 113,00

This book describes elements and applications of the graphic-oriented programming languages LAD (ladder diagram) and FBD (function block diagram) for use with both SIMATIC S7-300 and SIMATIC S7-400. It is aimed at all users of SIMATIC S7 programmable controllers. First-time users will be introduced to the field of programmable logic control whereas advanced users will learn about specific applications of SIMATIC S7 programmable controllers.